USING PHYSICS

GADGETS & GIZMOS

GRADES 9–12

PHENOMENON-BASED LEARNING

Matthew Bobrowsky

Mikko Korhonen

Jukka Kohtamäki

USING PHYSICS

GADGETS & GIZMOS
GRADES 9–12
PHENOMENON-BASED LEARNING

National Science Teachers Association

Arlington, Virginia

National Science Teachers Association

Claire Reinburg, Director
Wendy Rubin, Managing Editor
Andrew Cooke, Senior Editor
Amanda O'Brien, Associate Editor
Amy America, Book Acquisitions Coordinator

ART AND DESIGN
Will Thomas Jr., Director
Joe Butera, Senior Graphic Designer, cover and
  interior design

PRINTING AND PRODUCTION
Catherine Lorrain, Director

NATIONAL SCIENCE TEACHERS ASSOCIATION
David L. Evans, Executive Director
David Beacom, Publisher

1840 Wilson Blvd., Arlington, VA 22201
www.nsta.org/store
For customer service inquiries, please call 800-277-5300.

Cataloging-in-Publication Data are available from the Library of Congress.
ISBN: 978-1-936959-36-5
e-ISBN: 978-1-938946-62-2

CONTENTS

ABOUT THE AUTHORS   X

AN INTRODUCTION TO PHENOMENON-BASED LEARNING   XIII

To the Student   xiii
To the Teacher   xiii
PBL in Finland   xvi
Authors' Use of Gadgets and Gizmos   xvi
Safety Notes   xvii
References   xvii

1 PRESSURE AND FORCE   1

	Exploration	Analysis
Pressure Power	2	7
It's a Hold-Up!	3	8
Prism Pressure	4	9
Pressure Globe	5	10
Water Rocket	6	10

Web Resources	11
Relevant Standards	12

2 LAWS OF THERMODYNAMICS   15

	Exploration	Analysis
Meltdown	16	22
Pressure Plunger	18	23
Fire Syringe	19	24
Drinking Bird	20	25
Running Hot and Cold	21	26

Web Resources	27
Relevant Standards	28

3 ENERGY   31

	Exploration	Analysis
Energy on Wheels	32	40
Dancing Disc	33	41
Hot Shot	34	42
Radiant Rotation	35	42
Magnetic Accelerator	36	43
Happy / Unhappy Balls	37	43
Dropper Popper	38	44
Astroblaster	39	44

Web Resources   45
Relevant Standards   46

4 VISIBLE LIGHT AND COLORS   51

	Exploration	Analysis
Spectroscope	52	55
Adding and Subtracting Colors	53	56
Combining Colors	54	57

Web Resources   58
Relevant Standards   59

5   RESONANCE   63

	Exploration	Analysis
Standing Wave	64	69
Boomwhackers	65	70
Singing Rods	66	71
Music Box	67	71
Sound Pipe	68	72

Web Resources	73
Relevant Standards	74

6   BUOYANCY   79

	Exploration	Analysis
Cartesian Diver	80	86
Galileo's Thermometer	81	86
Mix It Up	82	86
Solar Bag	83	86
Rock the Boat	84	87

Web Resources	88
Relevant Standards	89

7   TWO-DIMENSIONAL MOTION   91

	Exploration	Analysis
Race to the Bottom	92	97
Ejector Seat	93	97
Need for Speed	94	98
Launch and Land	96	99

Web Resources	100
Relevant Standards	101

8

ANGULAR MOMENTUM 105

	Exploration	Analysis
Speed Spinning	106	111
Propeller Puzzle	107	112
Power Ball	108	112
Wobbly Top	109	113
Perpetual Top	110	114

Web Resources	115
Relevant Standards	116

9

MAGNETISM 119

	Exploration	Analysis
Point the Way	120	125
Magnets in Motion	121	126
All Encompassing	122	127
World's Simplest Motor	123	127
Levitron	124	128

Web Resources	129
Relevant Standards	130

10

ELECTROMAGNETIC INDUCTION 133

	Exploration	Analysis
Eddy Oddity	134	137
Shake & Shine	135	138
Crank-a-Current	136	138

Web Resources	139
Relevant Standards	140

		11	MORE FUN	**143**

	Exploration	Analysis
Wind Bag	144	150
Doppler Ball	145	151
Mirage	146	152
Fun Fly Stick	147	152
Energy Ball	148	153
Flying Pig	149	154

Web Resources	155
Relevant Standards	156

APPENDIX

How to Order the Gadgets and Gizmos	159

CREDITS	**161**
INDEX	**163**

ABOUT THE AUTHORS

MATTHEW BOBROWSKY, PHD

Dr. Matt Bobrowsky has been involved in scientific research and science education for several decades. He served for four years at the University of Maryland as director of the Physics Lecture Demonstration Facility—a collection of over 1,600 science demonstrations. Also at the University of Maryland, Matt was selected as a faculty mentor for the Fulbright Distinguished International Teachers Program, where he met Mikko Korhonen.

Matt's teaching is always innovative because he uses pedagogical techniques that are based on current science education research and known to be effective. Matt has taught physics, astronomy, and astrobiology both in the classroom and online. He has written K–12 science curricula and serves on the Science Advisory Committee for the Howard County Public School System in Maryland. Matt has conducted countless professional development workshops for science teachers and special presentations for students, speaking on a variety of topics beyond physics, such as the scale of the universe, life in the universe, misconceptions about science among students and the public, the process of science, and science versus pseudoscience. He is often asked to be an after-dinner speaker or keynote speaker at special events. Matt is a "Nifty Fifty" speaker for the USA Science & Engineering Festival and a Shapley Lecturer for the American Astronomical Society. Matt has received a number of awards for teaching excellence from the University of Maryland, including the Stanley J. Drazek Teaching Excellence Award and the Board of Regents' Faculty Award for Excellence in Teaching.

In his research, Matt has been involved in both theoretical and observational astronomy. He developed computer models of planetary nebulae—clouds of gas expanding outward from aging stars—and has observed them with telescopes on the ground as well as with the Hubble Space Telescope. One of the planetary nebulae that Matt investigated is the Stingray Nebula, which he discovered using Hubble.

MIKKO KORHONEN

Mikko Korhonen obtained a master's degree from Tampere Technical University in Finland, where he studied physics, mathematics, and pedagogics. Since then, he has been teaching physics, mathematics, and computer science at various schools in Finland. He has also developed a number of educational programs that brought some of his students to top scientific facilities in the world, including the Nordic Optical Telescope (NOT) observatory in La Palma, Spain, the CERN laboratory at the Franco-Swiss border, and the LATMOS laboratory in France. Most recently, some of his students have attended the Transatlantic Science School, which Mikko founded.

Mikko has written numerous other educational publications, including a book of physics experiments, manuals of physics problems with answers, an article on mathematics and logic for computer science, and two books with Jukka Kohtamäki on using toys to teach physics, one at the middle school and one at the high school level. (This book is an adaptation of the Finnish version of that high school book.)

Mikko has obtained numerous grants for his school and students, including those from the NOT science school and the Viksu science competition, as well as individual grants from the Finnish National Board of Education and the Technology Industries of Finland Centennial Foundation, and grants for his "physics toys" project. His students are also award winners in the Finnish National Science Competition. Mikko received one of the Fulbright Distinguished Awards in Teaching, which brought him to the University of Maryland, where he worked with Matt Bobrowsky. Most recently, Mikko received the Distinguished Science Teacher Award in 2013 from the Technology Industries of Finland Centennial Foundation.

JUKKA KOHTAMÄKI

Jukka Kohtamäki obtained his master of science from Tampere University of Technology in Finland and since then has been teaching grades 5–9 at the Rantakylä Comprehensive School, one of the largest comprehensive schools in Finland. Jukka has participated in long-term professional development teaching projects and projects involving the use of technology in learning, as well as workshops that he and Mikko Korhonen conducted for Finnish science teachers. His writing includes teaching materials for physics and computer science, and he has written two books with Mikko on using toys to teach physics, one at the middle school and one at the high school level. (This book is an adaptation of the Finnish version of that high school book.)

Jukka is a member of the group under the National Board of Education that is writing the next physics curriculum in Finland. He is also participating in writing curricula in chemistry and natural science (which is a combination of biology, geology, physics, chemistry, and health education). His goals are to get students engaged in lessons, to have them work hands on and minds on, to encourage creativity in finding solutions, and to get students to discuss natural phenomena using the "language of physics." In 2013, Jukka received the Distinguished Science Teacher Award from the Technology Industries of Finland Centennial Foundation.

"The most beautiful thing
we can experience is the
mysterious. It is the source
of all true art and science."

— Albert Einstein

AN INTRODUCTION TO PHENOMENON-BASED LEARNING

TO THE STUDENT

In 1931 Albert Einstein wrote, "The most beautiful thing we can experience is the mysterious. It is the source of all true art and science." Keep this in mind as we introduce you to phenomenon-based learning, a learning approach in which you start by observing a natural phenomenon—in some cases just a simple toy—and then build scientific models and theories based on your observations.

In science, there are many phenomena that are difficult to understand at first. Most physics books are written so that the theory—that is, the mathematical description—comes first, and demonstrations and applications are presented only afterward. In this book, by contrast, the goal is for you to first watch something happen and then to become curious enough to find out why. You will experiment with some simple gizmos and think about them from different perspectives. Developing a complete understanding of a concept might take a number of steps, with each step providing a deeper understanding of the topic. In some cases, you will need to do further research on your own to understand certain terms and concepts. Like real scientists, you can also get help from (and provide help to) collaborators. This book's approach to learning is based on curiosity and creativity—a fun way to learn!

TO THE TEACHER

The pedagogical approach in this book is called phenomenon-based learning (PBL), meaning learning is built on observations of real-world phenomena—in this case of some fun toys or gadgets. The method also uses peer instruction, which research has shown results in more learning than traditional lectures (Champagne, Gunstone, and Klopfer 1985; Crouch and Mazur 2001; Chi and Roscoe 2002). In the PBL approach, students work and explore in groups: exercises are done in groups, and students' conclusions are also drawn in groups. The teacher guides and encourages the groups and, at the end, verifies the conclusions. With the PBL strategy, the concepts and the phenomena are approached from different angles, each adding a piece to the puzzle with the goal of developing a picture correctly portraying the real situation.

The activities in this book can be used for various purposes. Ideas for how to approach the phenomena can be found in the introductions to the chapters. The introductions and the questions can be used as the basis for discussions with the groups before the students use the gizmos, that is, as a motivational tool. For example, you can ask where we see or observe the phenomenon in everyday life, what the students know about the matter prior to conducting the activities, and so on.

PBL is not so much a teaching method as it is a route to grasping the big picture. It contains some elements that you may have seen in inquiry-based, problem-based, or project-based learning, combined with hands-on activities. In traditional physics teaching, it's common to

AN INTRODUCTION TO PHENOMENON-BASED LEARNING

divide phenomena into small, separate parts and discuss them as though there is no connection among them (McNeil 2013; Verley 2008; Gray et al. 2008). In our PBL approach, we don't artificially create boundaries within phenomena. Rather, we try to look at physical phenomena very broadly.

PBL is different from project-based or problem-based learning. In project-based learning, the student is given a project that provides the context for learning. The problem with this is that the student is not necessarily working on the project out of curiosity but simply because they are required to by the teacher. To avoid having students view the project as a chore or just a problem that they have to solve, we employ PBL: The student's own curiosity becomes the drive for learning. The student explores not by trudging through a problem to get to the correct answer but by seeing an interesting phenomenon and wanting to understand what's going on. This works because interest and enthusiasm do not result from the content alone; they come from the students themselves as they discover more about a phenomenon. Personal experience with a phenomenon is always more interesting and memorable than a simple recitation of facts (Jones 2007; Lucas 1990; McDade 2013).

The goal in project-based learning is for the students to produce a product, presentation, or performance (Moursund 2013). PBL does not have that requirement; students simply enjoy exploring and discovering. This is the essence of science, and it is consistent with the philosophy of the *Next Generation Science Standards* (*NGSS*). Rather than simply memorizing facts that will

"Most of the time my students didn't need me; they were just excited about a connection or discovery they made and wanted to show me."

—Jamie Cohen (2014)

soon be forgotten, students are doing real science. They are engaged in collaboration, communication, and critical thinking. Through this, students obtain a deeper understanding of scientific knowledge and see a real-world application of that knowledge—exactly what was envisioned with the *NGSS*. This is why, at the end of each chapter, we provide a list of relevant standards from the *NGSS*, further emphasizing our focus on the core ideas and practices of science, not just the facts of science.

The objective of PBL is to get the students' brains working with some phenomenon and have them discussing it in groups. A gizmo's functions, in most cases, also make it possible for teachers to find common misconceptions that students may harbor. It is important to directly address misconceptions because they can be very persistent (Clement 1982, 1993; Nissani 1997). Often the only way to remove misconceptions is to have students work with the problem, experiment, think, and discuss, so that they can eventually experience for themselves that their preconception is not consistent with what they observe in the real world.

We must also keep in mind that students can't build up all the physics laws and concepts from scratch by themselves—unless you are lucky enough to have the next Newton or Einstein as one of your students. Students will definitely need some support and instruction. When doing experiments and learning from them, the students must have some qualitative discussions (to build concepts) and some quantitative work (to learn the measuring process and make useful calculations). Experience with both reveals the nature of physical science.

When you first look at this book, you might be struck by the fact that there is not very much textual material. That is intentional. The idea is to have more thinking by the students and less lecturing by the teacher. It is also important to note that the process of thinking and learning is not a race. To learn and really get the idea, students need to take time to think … and then think some more—so be sure to allow sufficient time for the cognitive processes to occur. For example, the very first experiment (Pressure Power) can be viewed in two seconds, but in order for students to think about the phenomenon and really get the idea, they need to discuss the physics with other group members, practice using the "language of physics," and internalize the physics involved—which might take 20 minutes. During this time, the students may also think of real-life situations in which the phenomenon plays a significant role, and these examples can be brought up later during discussions as an entire class.

You will also notice that there are no formal quizzes or rubrics included. There are other ways to evaluate students during activities such as these. First, note that the emphasis is not on getting the "right" answer. Teachers should not simply provide the answer or an easy way out—that would not allow students to learn how science really works. When looking at student answers, consider the following: Are the students basing their conclusions on evidence? Are they sharing their ideas with others in their group? Even if a student has the wrong idea, if she or he has evidential reasons for that idea, then that student has the right approach. After all members of a group are in agreement and tell you, the teacher, what they think is happening, you can express doubt or question the group's explanation, making the students describe their evidence and perhaps having

them discuss it further among themselves. Student participation as scientific investigators and their ability to give reasons for their explanations will be the key indicators that the students understand the process of science.

The PBL approach lends itself well to having students keep journals of their activities. Students should write about how they are conducting their experiment (which might differ from one group to another), ideas they have related to the phenomenon under investigation (including both correct and incorrect ideas), what experiments or observations showed the incorrect ideas to be wrong, answers to the questions supplied for each exploration, and what they learned as a result of the activity. Students might also want to make a video of the experiment. This can be used for later reference, as well as to show family and friends. Wouldn't it be great if we can get students talking about science outside the classroom?

A few of the questions asked of the students will be difficult to answer. Here again students get a feel for what it's like to be a real scientist exploring uncharted territory. A student might suggest an incorrect explanation. Other students in the group might offer a correction, or if no one does, perhaps further experimentation, along with guidance from the teacher, will lead the students on the right course. Like scientists, the students can do a literature search (usually a web search now) to see what others know about the phenomenon. Thus there are many ways for a misconception to get dispelled in a way that will result in more long-term understanding than if the students were simply told the answer. Guidance from the teacher could include providing some ideas about what to observe when doing the experiment or giving some examples from other situations in which the same phenomenon takes place. Although many incorrect ideas will not last long in group discussions, the teacher should actively monitor group discussions, ensuring that students do not get too far off track and are on their way to achieving increased understanding. We've provided an analysis of the physics behind each exploration to focus your instruction.

By exploring first and getting to a theoretical understanding later, students are working like real scientists. When scientists investigate a new phenomenon, they aren't presented with an explanation first—they have to figure it out. And that's what the students do in PBL. Real scientists extensively collaborate with one another; and that's exactly what the students do here as well—work in groups. Not all terms and concepts are extensively explained; that's not the purpose of this book. Again, like real scientists the students can look up information as needed in, for example, a traditional physics textbook. What we present here is the PBL approach, in which students explore first and are inspired to pursue creative approaches to answers—and have fun in the process!

PBL IN FINLAND

The Finnish educational system came into the spotlight after the Programme of International Student Assessment (PISA) showed that Finnish students were among the top in science literacy proficiency levels. Out of 74 countries, in 2009 Finland ranked 2nd in science and 3rd in reading. (The United States ranked 23rd and 17th, respectively.) In 2012, Finland ranked 5th in science and 6th in reading. (The United States ranked 28th and 24th, respectively.) Finland is now seen as a major international leader in education, and its performance has been especially notable for its

significant consistency across schools. No other country has so little variation in outcomes among schools, and the gap within schools between the top- and bottom-achieving students is quite small as well. Finnish schools seem to serve all students well, regardless of family background or socioeconomic status. Recently, U.S. educators and political leaders have even been traveling to Finland to learn the secret of their success.

The PBL approach is one that includes progressive inquiry, problem-based learning, project-based learning, and in Finland at least, other methods at the teachers' discretion. The idea is to teach bigger concepts and useful thinking skills rather than asking students to memorize everything in a textbook.

AUTHORS' USE OF GADGETS AND GIZMOS

One of the authors (M.B.) has been using gizmos as the basis of teaching for many years. He also uses them for illustrative purposes in public presentations and school programs. The other two authors (M.K. and J.K.) have been using PBL—and the materials in this book—to teach in Finland. Their approach is to present physics phenomena to students so that the students can build ideas and an understanding of the topic by themselves, in small groups. Students progress from thinking to understanding to explaining. For each phenomenon there are several different viewpoints from which the student can develop a big-picture understanding as a result of step-by-step exploration. The teacher serves only as a guide who leads the student in the right direction. PBL is an approach that is not only effective for learning but is also much more fun and interesting for both the teacher and the students.

SAFETY NOTES

Hands-on activities in science classrooms and laboratories help make learning science fun. In order to also make the activities safer, certain precautions must be followed based on legal standards and professional best practices. In this book, activities have appropriate and important safety notes listed that need to be followed for a safer experience. Prior to any activities taking place, students need to receive safety training, have a safety assessment, and sign and date (along with parents or guardians) a safety acknowledgement form.

REFERENCES

Champagne, A. B., R. F. Gunstone, and L. E. Klopfer. 1985. Effecting changes in cognitive structures among physics students. In *Cognitive structure and conceptual change*, ed. H. T. West and A. L. Pines, 163–187. Orlando, FL: Academic Press.

Chi, M. T. H. and R. D. Roscoe. 2002. The processes and challenges of conceptual change. In *Reconsidering conceptual change: Issues in theory and practice*, ed. M. Limón and L. Mason, 3–27. Boston: Kluwer Academic Publishers.

Clement, J. 1982. Students' preconceptions in introductory mechanics. *American Journal of Physics* 50 (1): 66–71.

Clement, J. 1993. Using bridging analogies and anchoring intuitions to deal with students' preconceptions in physics. *Journal of Research in Science Teaching* 30 (10): 1241–1257.

Cohen, J. 2014. 18 Ways to engage your students by teaching less and learning more with rap genius. *http://poetry.rapgenius.com/Mr-cohen-18-ways-to-engage-your-students-by-teaching-less-and-learning-more-with-rap-genius-lyrics*

Crouch, C. H., and E. Mazur. 2001. Peer instruction: Ten years of experience and results. *American Journal of Physics* 69 (9): 970–977.

Gray, K. E., W. K. Adams, C. E. Wieman, and K. K. Perkins. 2008. Students know what physicists believe, but they don't agree: A study using the CLASS survey. *Physics Review Special Topic–Physics Education Research* 4: 020106

Jones, L. 2007. *The student-centered classroom*. New York: Cambridge University Press. *www.cambridge.org/other_files/downloads/esl/booklets/Jones-Student-Centered.pdf*.

Lucas, A. F. 1990. Using psychological models to understand student motivation. In *The changing face of college teaching: New directions for teaching and learning*, no. 42, ed. M. D. Svinicki, 103–114. San Francisco: Jossey-Bass

McDade, M. 2013. Children learn better when they figure things out for themselves: Brandywine professor's research published in journal. *PennState News. http://news.psu.edu/story/265620/2013/02/21/society-and-culture/children-learn-better-when-they-figure-things-out*.

McNeil, L. E. Transforming introductory physics teaching at UNC-CH. University of North Carolina at Chapel Hill. *http://user.physics.unc.edu/~mcneil/physicsmanifesto.html*.

Moursund, D. 2013. Problem-based learning and project-based learning. University of Oregon. *http://pages.uoregon.edu/moursund/Math/pbl.htm*.

Nissani, M. 1997. Can the persistence of misconceptions be generalized and explained? *Journal of Thought* 32: 69–76. *www.is.wayne.edu/mnissani/pagepub/theory.htm*.

Verley, J. D. 2008. Physics graduate students' perceptions of the value of teaching. PhD diss., University of Wyoming. *http://udini.proquest.com/view/physics-graduate-students-goid:304450532*.

PRESSURE AND FORCE

In everyday life we often talk about pressure—high pressure and low pressure. We speak of the pressure in car tires, atmospheric pressure, pressure-sealed containers (such as pressure cookers), pressure chambers, and so on. These are just a few of the ways we encounter pressure in daily life.

Sometimes, we experience pressure more directly. While flying in airplanes or when visiting the Empire State Building our ears may "pop." Drinking bottles may bulge or flatten. Children are able to drink from juice boxes with straws. All of these phenomena are related to changes in pressure.

In this chapter, we will acquaint you with the forces caused by pressure. With hands-on activities, you will get to know what effects air pressure and related forces produce in everyday phenomena. Most importantly, you will come to understand what causes air pressure.

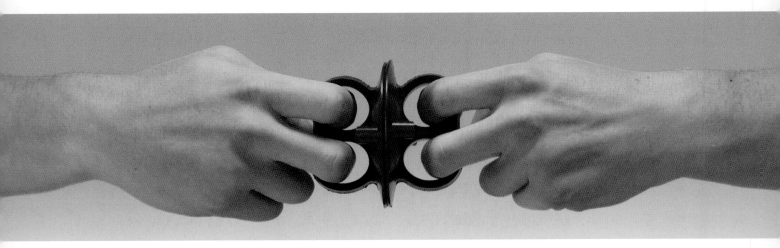

FIGURE 1.1: Atmospheric Pressure Cups

▼ SAFETY NOTE

Wear safety glasses or goggles.

PRESSURE POWER

We are used to normal air pressure, so we do not necessarily realize just how big a force normal atmospheric pressure can create. With the Atmospheric Pressure Cups (Figure 1.1), you create an amount of pressure that is smaller than normal air pressure. During the next demonstration, it is important to consider what forces pressure creates and how they affect the demonstration.

Procedure

1. Push the Atmospheric Pressure Cups together, squeezing out the air between them.

2. Try to separate the cups; brace yourself so you don't fall backward.

Questions

- Explain how the Atmospheric Pressure Cups work.

- What are the forces on one of the cups?

IT'S A HOLD-UP!

This demonstration works on the same principle as the Atmospheric Pressure Cups. With the Atmospheric Mat (Figure 1.2), we take a closer look at the forces of pressure and calculate how much weight you are able to lift using those forces.

Procedure

1. Attach a hook to the rubber mat.

2. Choose something to lift, such as a chair or small table. Make sure it is light enough—less than 20 kg—and has a flat surface.

3. Place the mat on the smooth surface of the table, and using the hook, slowly lift the table.

4. Place a tablecloth or a piece of cloth between the surface of the table and the mat, and slowly try to lift the mat.

5. Calculate the (theoretical) maximum weight you can lift with the mat.

6. Keep the table legs very close to the floor as you support the table with the hook and have someone add an additional downward force to the table.

Questions

- What force or forces are holding up the mat?

- What force or forces are holding up the table?

- What is the theoretical maximum weight you can lift with the Atmospheric Mat?

- If the table came off of the mat, explain why there is a big difference between theory and reality.

- What changed when you added the cloth? Why?

▼
SAFETY NOTES

- Wear safety glasses or goggles.

- Use caution when lifting the chair or table. Keep your feet out of the way in case the chair or table falls to the floor.

FIGURE 1.2: Atmospheric Mat

▼
SAFETY NOTES

- Wear splash goggles and an apron.

- Immediately wipe up any splashed water to prevent a slip or fall hazard.

FIGURE 1.3: Hollow Prism

PRISM PRESSURE

Air pressure at sea level can be explained with the help of some of the concepts of hydrostatic gas pressure. Hydrostatic pressure also applies to liquids. For example, the hydrostatic pressure at the bottom of a column of water increases as the height of a column of water increases. (For columns of any liquids, the pressure will depend on both the height and the density of the liquid.) There are many forces in play when using the Hollow Prism (Figure 1.3), and here you can investigate some of those forces and their effects.

Procedure

1. Fill the prism with water, removing all air bubbles, and close the cap.

2. Turn the prism upside down and remove the cap.

3. *Test 1*: Slide a narrow stick (e.g., a mechanical pencil lead) into the prism when the prism is upside down.

4. *Test 2*: Slowly insert a thin straw into the prism when the prism is upside down.

5. Repeat the tests, but this time add a little dishwashing detergent to the water.

Questions

- Does the water flow out of the prism? Why or why not?

- During which step did something happen? Explain.

- Experiment with inserting the straw to different depths and see what happens. Report your results.

PRESSURE GLOBE

The first three explorations focused on pressure and its effects. In the following demonstrations using the Pressure Globe (Figure 1.4), these effects are easy to see.

Procedure

1. Remove the plug below the globe and then blow air into the balloon at the top of the globe.

2. Keep the opening of the balloon closed and, without letting air out of the balloon, put the plug back in place.

3. Release the opening of the balloon.

4. Now, with the plug in place, blow more air into the balloon.

5. Remove the plug.

Questions

*   What happens to the balloon inside the glass when the plug is in place?

*   Why is it hard to blow more air into the balloon?

*   Repeat the test, but this time go to a sink and add about 1 cup of water to the balloon (with the plug in place). Then, over the sink or outdoors, remove the plug.

*   What happens, and what is the force that causes it to happen?

FIGURE 1.4: Pressure Globe

SAFETY NOTES

*   Wear splash goggles and an apron.

*   Immediately wipe up any splashed water to prevent a slip or fall hazard.

WATER ROCKET

You'll need an open field for this experiment.
The altitude of the rocket's flight from the Super
Bottle Rocket Launcher (Figure 1.5) may surprise
you, so make sure that the launch area is large
enough (a clear area with at least a 10-m radius).
The water rocket can be used to observe the
effects that pressure forces can achieve.

FIGURE 1.5: Super Bottle Rocket Launcher

Procedure

1. Place the holder firmly into the ground.

2. Add a half-full bottle of water.

3. Place the bottle on the rack, and attach the pin.

4. Using a pump with a pressure gauge, pump the
    bottle to about 5 bars of pressure.
    (5 bars = 73 PSI = 500 kPa)

5. Launch the rocket by removing the pin with a
    piece of string.

Questions

*   What happens to the water—and the air—inside the
    bottle?

*   Explain why the bottle rises into the air.

*   Explain why some fog appears in the bottle.

▼ SAFETY NOTES

*   Wear safety glasses or
    goggles.

*   Make sure the trajectory
    path of the water rocket
    is clear of observers,
    fragile equipment,
    and so on, to prevent
    damage or injury.

Analysis 　1A

FIGURE 1.6: Atmospheric Pressure Cups

PRESSURE POWER

Gas contains an enormous number of molecules that, because of their thermal motion, are moving in random directions at varying speeds. Gas molecules will collide with the wall of whatever container they're in, which in this case is the wall of the Atmospheric Pressure Cups (Figure 1.6). These collisions create impulses on the wall. A certain number of collisions are occurring on every unit area all the time. The gas pressure–induced force may be thought of simply as the forces caused by these impulses, with the average direction being toward the container wall. Pressure (p) is defined as the force (F) per unit area (A): $p = F / A$. Therefore the gas-induced force can be written as $F = pA$.

The gas pressure inside a container creates a force that pushes out the walls of the container. Outside a container, gas pressure creates a force that pushes the walls inward. If the air pressure outside is equal to the air pressure inside, the pressure forces are balanced.

Between the pressure cups, there is space in which the amount of air cannot change. Pulling the pressure cups apart makes the volume of the air space increase, and because the number of gas molecules remains constant, the air pressure inside lowers. At that point, the air pressure outside is greater than the air pressure in the air space between the cups, so there is a net force pushing the cups toward each other.

The pressure cup forces include

- force from you pulling the cups apart,

- force caused by the external air pressure (compressing the cups together), and

- force caused by air pressure inside the cups (pushing the cups apart).

If the cups are not moving, the sum of all the forces is zero. Therefore, the outside air pressure compresses the pressure cups toward each other.

IT'S A HOLD-UP!

The Atmospheric Mat (Figure 1.7) works on the same principle as the Atmospheric Pressure Cups. When the table is lifted with the mat, an air space is formed between the table and the mat. The forces on the mat are as follows:

- Hook "support force" (i.e., you pulling up on the hook)

- Force created by air pressure under the mat (between the table and the mat)—pushing up

- Force created by air pressure above the mat—pushing down

- Gravitational force on the mat (i.e., its own weight)—pulling down

FIGURE 1.7: Atmospheric Mat

The force from the air above the mat is greater than the force under the mat (between the mat and the table). The sum of all the forces is zero. That is, the sum of the hook support force and the upward force from the air pressure under the mat (which pushes in all directions equally) is equal to the downward force from the air pressure pushing on the top of the mat plus the (small) weight of the mat.

Pressure is defined as $p = F / A$ (where F is force and A is area), or $F = pA$. If the Atmospheric Mat had a complete vacuum under it (which is only possible in theory), the air pressure difference between above and below the mat would be equal to the atmospheric pressure $p = 101$ kPa. The mat has an area of 0.2 m × 0.2 m = 0.04 m².

$F = pA = (101 \text{ kPa}) (10^3 \text{ Pa/kPa}) (0.04 \text{ m}^2)$

$= 4 \times 10^3 \text{ N}$

That's equivalent to the weight of 400 kg.

This theoretical maximum value is different from the actual value because a complete vacuum is not achieved under the Atmospheric Mat. Also, the area where the force applies is somewhat smaller than the theoretical maximum.

The vertical forces on the table include its weight (i.e., gravitational force from the Earth), air pressure under the table pushing up, and (lower) air pressure under the mat pushing down (see the Analysis of the Atmospheric Pressure Cups for more on this). Since the table is not accelerating, the net force on it must be zero. Thus, the upward air pressure force is equal to the sum of the downward, lower air pressure force under the mat and the gravitational force. (This assumes that the mat covers the entire top of the table; otherwise there is the additional downward force of air pressure on the parts of the table top beyond the mat.)

FIGURE 1.8: Hollow Prism

PRISM PRESSURE

The reason water doesn't drain out of the Hollow Prism (Figure 1.8) is rather interesting. The prism has a hole on one end, which can be covered with a cap. If the hole were larger, then water would start to drain out of one side of the hole while air formed a bubble and entered another side of the hole. However, the hole in the prism is so small that the surface tension, which acts parallel to the surface, keeps the water surface from deforming enough to form a bubble over one part of the surface and a drip over the other.

A piece of pencil lead (narrow stick) does not affect the pressure difference or break the surface tension of the water, so the lead does not cause the water to drain out. However, a straw can make a big difference. For any container of water, the hydrostatic pressure will be greatest at the bottom. The air inside the straw has the same pressure as the outside air, but the pressure is not great enough to push air into the water inside the prism at the bottom. When the top of the straw rises inside the prism, the hydrostatic pressure of the water at the top of the straw diminishes. At a certain height, the external pressure is high enough that air starts to flow into the prism. At the same time water, displaced by air, flows out of the prism. The straw also helps by providing one channel for the water to flow out and another one for the air to flow in. Students might notice air going up the straw and water flowing down the outside of the straw.

Detergent reduces the surface tension of the water, so the water drains more easily out of the prism.

FIGURE 1.9: Pressure Globe

PRESSURE GLOBE

After you inflate the balloon inside the Pressure Globe (Figure 1.9) and replace the plug, the balloon does not deflate. The elastic forces in the balloon's material and air pressure between the globe and the balloon try to compress the balloon, but air pressure from the outside, pushing into the balloon, prevents the compression. The outside air pressure pushing into the balloon is greater than the air pressure inside the globe.

Next, you blow into the balloon while the plug is in place. Blowing into the balloon creates a higher pressure, but the balloon can't expand very much because of the air trapped in the globe and because the globe can't expand. After blowing, the pressure in the balloon is the same as the outside air pressure because the balloon's mouth is open.

Finally, you remove the plug. Once this is done, the inside of the globe has the same pressure as the surrounding atmospheric pressure. Therefore the pressure forces on the inside and outside of the balloon cancel each other out, and the elastic force in the balloon material pushes air (or water) out from the globe.

WATER ROCKET

In the bottle of water on the Super Bottle Rocket Launcher (Figure 1.10), the pressure is higher than the normal atmospheric pressure. When the rocket is launched, the pressure stabilizes to normal atmospheric pressure as water is pushed rapidly out of the bottle. The air pressure pushes the water downward and the bottle upward. The momentum of the system is conserved, so the momentum of the water moving downward has the same magnitude as the momentum of the rocket moving upward.

FIGURE 1.10: Super Bottle Rocket Launcher

The rocket achieves a high speed due to its lightness (low mass) compared with the water mass.

When the rocket lands, fog can be seen inside the bottle. This is the result of the sudden decrease in pressure and temperature. When the rocket is launched, the pressure starts out high but then drops rapidly to the normal level of air pressure. Since the pressure change occurs so rapidly, the temperature differences do not have time to equalize, and the temperature is reduced. (This is called an *adiabatic* process, which is discussed more in Chapter 2.) As the temperature decreases, the relative humidity inside the bottle increases, although the absolute amount of water vapor will remain the same. A cloud forms inside the bottle when the relative humidity becomes 100% and the dew point (when vapor turns to liquid) is reached.

Web Resources

Exercises dealing with fluid pressure and depth.
www.grc.nasa.gov/www/K-12/WindTunnel/Activities/fluid_pressure.html

A Pitot tube activity.
www.grc.nasa.gov/www/K-12/BGA/Melissa/pitot_act.htm

Equation of state problems.
www.grc.nasa.gov/www/K-12/BGA/Corrine/equation_of_state_act.htm

An activity on the relationships among altitude, air density, temperature, and pressure.
www.grc.nasa.gov/www/K-12/problems/Jim_Rinella/AltitudevsDensity_act.htm

An activity on the relationship between air pressure and temperature.
www.grc.nasa.gov/www/K-12/Missions/Rhonna/pre_act.htm

Pascal's principle and fluid pressure.
www.grc.nasa.gov/www/K-12/WindTunnel/Activities/Pascals_principle.html

Control a piston in a chamber to explore relationships among pressure, temperature, density, and volume.
http://jersey.uoregon.edu/vlab/Piston/

A simulation to investigate how pressure changes in air and water.
http://phet.colorado.edu/en/simulation/under-pressure

Questions to go with the PhET "Under Pressure" simulation.
http://phet.colorado.edu/en/contributions/view/3611

A fluid pressure activity.
http://phet.colorado.edu/en/simulation/fluid-pressure-and-flow

An activity to explore pressure at depth, Pascal's principle, hydraulic pistons, and absolute versus gauge pressure.
http://phet.colorado.edu/en/contributions/view/3619

Relevant Standards

Note: The Next Generation Science Standards *can be viewed online at* www.nextgenscience.org/next-generation-science-standards.

PERFORMANCE EXPECTATIONS

HS-PS2-2

Use mathematical representations to support the claim that the total momentum of a system of objects is conserved when there is no net force on the system. [Clarification Statement: Emphasis is on the quantitative conservation of momentum in interactions and the qualitative meaning of this principle.]

SCIENCE AND ENGINEERING PRACTICES

- Plan and conduct an investigation individually and collaboratively to produce data to serve as the basis for evidence, and in the design: decide on types, how much, and accuracy of data needed to produce reliable measurements and consider limitations on the precision of the data (e.g., number of trials, cost, risk, time), and refine the design accordingly.

- Analyze data using tools, technologies, and/or models (e.g., computational, mathematical) in order to make valid and reliable scientific claims or determine an optimal design solution.

- Constructing explanations and designing solutions in 9–12 builds on K–8 experiences and progresses to explanations and designs that are supported by multiple and independent student-generated sources of evidence consistent with scientific ideas, principles, and theories.

- Apply scientific ideas to solve a design problem, taking into account possible unanticipated effects.

CONNECTIONS TO NATURE OF SCIENCE

Science Models, Laws, Mechanisms, and Theories Explain Natural Phenomena

- Theories and laws provide explanations in science.

- Laws are statements or descriptions of the relationships among observable phenomena.

DISCIPLINARY CORE IDEAS

PS1.A: Structure and Properties of Matter

- Matter of any type can be subdivided into particles that are too small to see, but even then the matter still exists and can be detected by other means. A model showing that gases are made from matter particles that are too small to see and are moving freely around in space can explain many observations, including the inflation and shape of a balloon and the effects of air on larger particles or objects.

CROSSCUTTING CONCEPTS

Patterns

- Different patterns may be observed at each of the scales at which a system is studied and can provide evidence for causality in explanations of phenomena.

Cause and Effect

- Empirical evidence is required to differentiate between cause and correlation and make claims about specific causes and effects.

- Systems can be designed to cause a desired effect.

Systems and System Models

- When investigating or describing a system, the boundaries and initial conditions of the system need to be defined.

2

LAWS OF
THERMODYNAMICS

Thermodynamics is the physics of heat, mechanical energy or work, and the conversion of one into the other. Classical thermodynamics was developed after the invention of the steam engine in the early 1800s.

All systems follow the laws of thermodynamics. Since almost all energy production is based on thermodynamics, it is important to understand the laws of thermodynamics. The general rules of thermodynamics explain wind, ocean currents, and internal combustion engines.

FIGURE 2.1: Ice Melting Blocks

▼
SAFETY NOTE

Immediately wipe up any splashed water to prevent a slip or fall hazard.

MELTDOWN

The following experiment is an example of the second law of thermodynamics: isolated systems move toward *thermodynamic equilibrium*, which means that temperature differences tend to even out. In this case, two similar-looking blocks (Figure 2.1) are used to observe a phenomenon in physics called *heat conduction*.

PART 1

Procedure

1. Touch both blocks with your fingers at the same time to compare their temperatures.

2. Place an ice cube on each block.

3. Watch how the ice melts.

Questions

• Why is there a difference between how the ice melts on the two blocks?

• Did one of the two blocks start out warmer? If so, which one? Why?

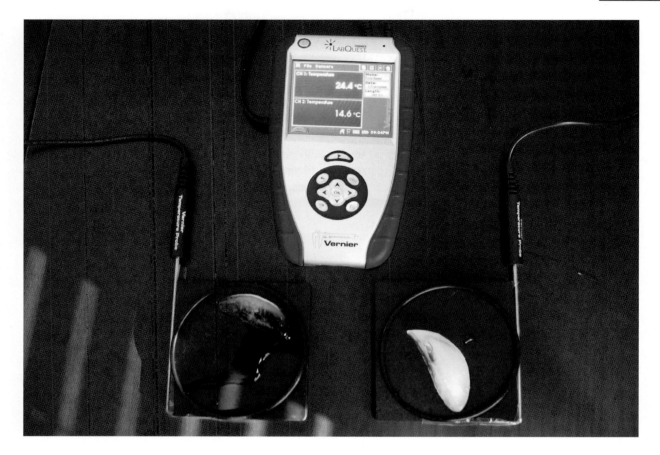

FIGURE 2.2: Temperature sensors

PART 2

Procedure

1. Attach the blocks to the temperature sensors (Figure 2.2).

2. Watch for changes in temperature while the ice is melting.

3. After the ice has melted, touch the table under the blocks with your hand to see if there is a temperature difference.

Questions

- Where does the energy come from to melt the ice?

- How does the table feel under each block? Explain.

▼
SAFETY NOTES

- Wear safety splash goggles.

- Immediately wipe up any splashed water to prevent a slip or fall hazard.

FIGURE 2.3: Elasticity of Gases Demo

▼
SAFETY NOTE

Wear safety glasses or goggles.

PRESSURE PLUNGER

The Elasticity of Gases Demo (Figure 2.3) allows you to explore the relationship between pressure, *P*, and volume, *V*, while the temperature is held constant. This is also known as Boyle's law. For the first and second laws of thermodynamics, it is important to know how the gas pressure changes as a function of volume.

Procedure

1. Measure the mass of the piston and upper woodblock. (This is the mass that compresses the air.)

2. Measure the inside diameter and calculate the cross-sectional area of the tube.

3. Place the piston in the tube and push the tube down to some volume, for example 40 ml.

4. Close the tube with a cap.

5. Note the volume, and calculate the pressure inside the tube. Remember to add the atmospheric pressure to the pressure caused by the wood block.

6. Collect the pressure and volume data.

7. Add the first weight (500 g) to the top of the wooden block, and calculate the new *P-V* values.

8. Add more weights to the block so that you get a good amount of data. Eight to ten measurements with different weights will be enough.

9. Draw a *P-V* graph based on your data.

10. If time permits, repeat the measurements with the same weights outdoors when the air temperature is much lower. Mark the values on the same graph as above.

Questions

- How are *P* and *V* related when the temperature is constant?

- In Step 10, how does the gas temperature affect the *P-V* graph?

FIRE SYRINGE

Rapidly increasing the pressure of a gas will raise the temperature of the gas. The Fire Syringe (Figure 2.4) shows this in a dramatic way. The faster the pressure increases, the faster the temperature increases. The operation of a diesel engine is based on this phenomenon.

Procedure

1. Place a very small amount of cotton (not much bigger than a grain of rice) at the bottom of the tube.

2. Screw the cap on to the tube with the piston raised.

3. Turn off the lights.

4. Sharply (as fast as possible) push the piston down.

Questions

*   Explain what happened.

*   Sketch a graph using (P, V) coordinates showing what happened.

*   Find out what an adiabatic process is and what a Carnot engine is. Discuss with the class and write about them in your science journal.

FIGURE 2.4: Fire Syringe

▼
SAFETY NOTES

*   Wear safety glasses or goggles.

*   Use caution when moving in the darkened room to prevent injury.

FIGURE 2.5: Drinking Bird

DRINKING BIRD

The Drinking Bird (Figure 2.5) is a well-known toy that
demonstrates several principles of thermodynamics.
To understand the physics of the gas in the Drinking
Bird, one must take into account effects of temperature
and the basic equations of mechanics.

Procedure

1. Soak the bird's head with cold water.

2. Place a water cup in front of the bird and blow on its head.

3. If the bird does not begin to "peck" at the water, adjust
   the bird's center of gravity so that the bird is just
   barely leaning against the water cup.

4. Warm the bird's head and then the bird's bottom with
   your hands. Watch the level of the liquid in the bird as
   you do this.

Questions

- Explain the physics of the Drinking Bird.

- In step 4, what happens? Why?

RUNNING HOT AND COLD

The Reversible Thermoelectric Demo (Figure 2.6) allows you to explore how a thermal (or heat) engine works. The thermal engine is a machine that converts thermal energy to kinetic energy—the energy of motion. A heat pump creates a difference in temperature with electrical energy.

PART 1

Procedure

1. Put warm water in one cup and cold water in another.

2. Place the aluminum legs in the cups so that one leg is in hot water and one is in cold water.

3. Make sure that the switch is turned to the right when viewed from the front.

4. Wait a few minutes, and then gently spin the propeller.

Questions

- Where did the fan get its energy to spin?

- What happens to the temperatures of the cups over time?

- Draw an energy flow diagram for this thermal power machine. (Use arrows on the photo of this gadget to depict the flow of energy from one part to another.)

▼
SAFETY NOTES

- Wear safety splash goggles.

- Immediately wipe up any splashed water to prevent a slip or fall hazard.

PART 2

Procedure

1. Put water with the same temperature in both cups.

2. Connect the power supply (6V DC) to the electric motor.

FIGURE 2.6: Reversible Thermoelectric Demo

3. Make sure that the switch is turned to the left when viewed from the front.

4. Monitor the temperature of the water with the temperature sensors.

Questions

- What happens to the temperature of the water in the cups?

- Draw the energy flow diagram of the heat pump.

▼
SAFETY NOTES

- Wear safety splash goggles.

- Immediately wipe up any splashed water to prevent a slip or fall hazard.

- Connect the power supply to a ground-fault circuit interrupter–protected circuit to prevent shock hazards.

Analysis

FIGURE 2.7: Ice Melting Blocks

MELTDOWN

In this first demonstration of thermodynamics, blocks of different materials (Figure 2.7) are used to investigate heat conduction. Heat is the random movement—or vibration—of particles in some material. The more the atoms and molecules vibrate, the warmer the substance is. Heat conduction is the transfer of vibrations from one part of the material to the next part. When an object is in contact with some material that is warmer, the vibrations are transferred from the warmer material to the cooler material.

In thermal insulators, atomic vibrations are not easily transferred through the material. "Meltdown" shows an example of this with a foam block. One reason that insulators sometimes do not conduct the vibrations very well is that they contain very small empty spaces filled with air, which slows down the heat conduction.

The second block in the demonstration is made of aluminum. In addition to the vibration of molecules in the metal, there are electrons that also transfer heat. Metals allow the negatively charged electrons to move freely among the positively charged metal ions and can transfer heat energy in metals quickly. The aluminum block feels cold at first because it rapidly conducts heat from your hand. In the case of the foam, your fingers do not feel cold because heat from your hand is not conducted away quickly.

The ice cube placed on top of the (aluminum) thermal conductor melts because the conductor transfers heat through the block to the ice cube. Heat is transferred as long as there is a temperature difference. The ice cube on top of the (foam) insulator does not melt as fast because the heat does not effectively transfer through the block to the ice. The stationary air around the ice cube also forms an insulating layer, which slows the transfer of heat. The ice cube would melt faster if the air around it were moving.

One of the consequences of the second law of thermodynamics is that in an isolated system, temperature differences are equalized.

FIGURE 2.8: Elasticity of Gases Demo

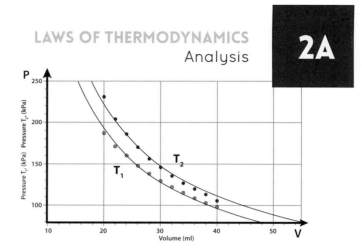

FIGURE 2.9: A graph showing both a cold-gas and a hot-gas isotherm

PRESSURE PLUNGER

Boyle's law describes how as the volume of gas decreases, the pressure increases (and vice versa). You can see this in action when you determine the volume and pressure of gas inside a tube as you press down on a plunger in the Elasticity of Gases Demo (Figure 2.8).

If the temperature is kept constant then, as the volume decreases, the pressure increases. This makes sense if you think about what's happening with the molecules. As you squeeze the gas into a smaller volume, the molecules are closer together, and collisions with the walls of the container occur more frequently. It is these collisions that cause gas to exert pressure. More frequent collisions means higher pressure.

The pressure and volume are inversely related: PV = constant, or P = constant/V. That's what the graph should show. (See Figure 2.9.)

If the temperature is kept constant, the graph of pressure versus volume is called an isotherm. In the pressure-volume (*P-V*) graph, the isotherm of a hot gas (T_2) is therefore above the cold gas (T_1), as shown in Figure 2.9. This explains the difference in results between the indoor and the outdoor experiments.

Notice that as the volume increases, the pressure decreases, exactly as you'd expect if the product of P times V equals a constant. (Do the curves look like hyperbolas?) Since P = constant/V, if you were to graph P versus $1/V$, you should get something close to a straight line, as shown in Figure 2.10.

Summary: Heat is the random movement of particles in some material. If the pressure is constant (which is usually the case when gas is free to expand), then the more heat there is (that is, the more random movement of particles), the greater the volume of the system. Thus, heating air in an open environment causes the air to expand. In this experiment, the plunger was kept on the tube, so the pressure could not remain constant as the volume or temperature changed. In a confined space (constant volume), a higher temperature results in a higher pressure. If you slowly squeeze the air into a smaller space (decreased volume), the pressure will increase. If you squeeze the air very rapidly, both the pressure and temperature will increase, which is what happens in the next exploration—the "Fire Syringe."

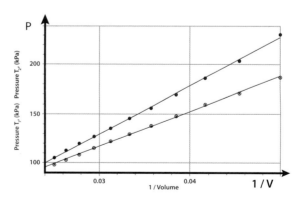

FIGURE 2.10: If you graph *P* versus *1 / V*, you get a straight line

FIRE SYRINGE

This experiment looks again at the relationship between volume, pressure, and temperature in respect to gases. When depressing the piston in the Fire Syringe (Figure 2.11), work is done on the gas, and its temperature rises enough to ignite a small piece of cotton. The molecules of air collide with the top, bottom, and sides of the glass tube. Depressing the piston increases the speed of the molecules. As the piston goes down, the volume of the gas in the tube decreases and the pressure increases. The relation between pressure and volume can be seen in a *P-V* graph.

The work done when depressing the piston and compressing the air in the tube is equal to the area under the *P-V* curve (Figure 2.12).

When the temperature is higher, the *P-V* curve is higher. As the piston goes down and the volume of gas decreases, the air molecules are repeatedly hit by the lowering piston; their kinetic energy increases, which means that the temperature of the air goes up.

Because the piston is depressed so rapidly, the temperatures of the air in the tube and the air in the room do not have time to equalize, and the *P-V* graph looks like the graph in Figure 2.12.

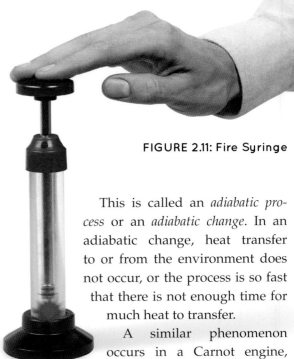

FIGURE 2.11: Fire Syringe

This is called an *adiabatic process* or an *adiabatic change*. In an adiabatic change, heat transfer to or from the environment does not occur, or the process is so fast that there is not enough time for much heat to transfer.

A similar phenomenon occurs in a Carnot engine, which is an ideal heat engine, meaning that it works without any energy loss. (This is actually impossible to have, but we can discuss it theoretically.) You can do a little research to find out how a Carnot engine works, or read more about it below.

Extension on the Carnot Engine

The operation of a Carnot engine can be described using a *P-V* diagram (Figure 2.13).

When going from point 1 to point 2, gas is compressed (as when depressing the piston in the Fire

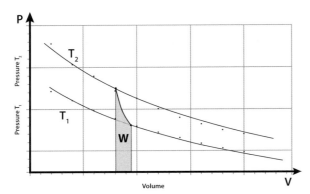

FIGURE 2.12: A *P-V* graph of the Fire Syringe exploration. The area under the *P-V* curve shows the work (*W*) done

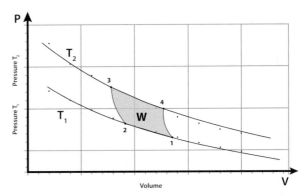

Figure 2.13: A *P-V* graph highlighting the work (*W*) done in the circuit 1-2-3-4

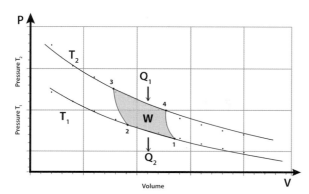

FIGURE 2.14: A *P-V* graph highlighting the work (*W*) done by a Carnot engine. *W* = *Q*$_1$ - *Q*$_2$.

Syringe). The gas's volume decreases, and its pressure increases. But unlike in the case of the Fire Syringe, the change is isothermal, which means that the temperature remains constant. The environment performs work on the system. (This work is the area under the curve from point 1 to point 2.) Since the temperature remains constant and the environment does work on the system, the gas radiates heat (an amount Q_2, as shown in Figure 2.14) into the environment.

When going from point 2 to point 3 in the Carnot engine, an adiabatic change occurs in which a rapid increase in gas pressure causes the gas temperature to rise rapidly—so fast that there is not enough time for heat to transfer to the environment.

When going from point 3 to point 4, the gas expands, and as the expanding gas cools down, it takes in heat from the environment (an amount Q_1). The engine performs work equal to the area under the curve from point 3 to point 4.

Going from point 4 to point 1 is an adiabatic process in which the gas pressure is reduced rapidly. There's not enough time for much heat to be transferred from the environment. Therefore, heat is lost to the environment between points 1 and 2 (Q_2), and heat is gained from the environment between points 3 and 4 (Q_1). The difference between Q_1 and Q_2 (actually the area in the circuit 1-2-3-4) is the work done by the Carnot engine. The efficiency of this engine is the work done divided by the amount of heat energy transferred into the system, Q_1.

DRINKING BIRD

The Drinking Bird (Figure 2.15) consists of two glass bulbs connected by a thin glass tube. One bulb forms the bird's head, and the other is the bird's lower body, which contains some liquid. At first, both bulbs have the same pressure and temperature, and the bird is in equilibrium.

FIGURE 2.15: Drinking Bird

When the bird's beak or head is moistened, the water begins to evaporate from the bird's head, making it cool off. As the temperature of the head drops, it gets closer to the dew point (the point at which vapor turns to liquid). When the temperature reaches the dew point, some of the vapor inside the head condenses into a liquid. The pressure then decreases. The lower pressure makes the fluid rise from the lower body along the connecting glass tube and changes the bird's center of mass. When the liquid reaches the bird's head, the bird tips over so that its beak goes into the water cup. As this happens, the liquid flows back down to the body, the pressures in the upper and the lower containers are equalized, and the bird swings upright. The bird then has some kinetic energy and swings back and forth like a pendulum. (A minor contribution to the cooling also comes from the bird's oscillations around the equilibrium position, causing water evaporation to speed up from the increased airflow past the bird's head.) The process repeats as long as there is enough water to keep the bird's head wet.

In step 4, you experimented with warming the bird's bottom, which you may have noticed causes the fluid to rise. In this case, the higher gas pressure pushes the fluid up the tube.

RUNNING HOT AND COLD

Based on the second law of thermodynamics, heat will flow from hot to cold (higher temperature to lower temperature), causing temperature differences to equalize. In the Reversible Thermoelectric Demo (Figure 2.16) energy is transferred through the aluminum feet from warm to cold water. Some of the thermal energy is changed into electrical energy, and the electrical energy runs a motor, which rotates the propeller.

When heat flows from hot to cold, energy can be released that can be used to do work, as shown in Figure 2.17.

If energy is added to the system, heat can be transferred from cold to hot (Figure 2.18). In this case, adding electrical power to the system makes two containers of water that start out at the same temperature end up with one cooler and the other warmer. Thermal energy is therefore transferred from one water container to another. This setup can function as a cooling machine, like a refrigerator.

This reversible thermoelectric demonstration is based on the Peltier and Seebeck effects. The Seebeck effect is the production of electricity from a

FIGURE 2.16: Reversible Thermoelectric Demo

temperature difference. Between the aluminum legs there is a thermocouple, which produces an electrical voltage due to the difference in temperatures between the two containers of water. The Seebeck effect has practical uses, for example, in the measurement of temperature.

The Peltier effect occurs when the current flowing through the thermocouple creates a temperature difference across the element. The Peltier effect is used for, among other things, cold-storage compartments for camping.

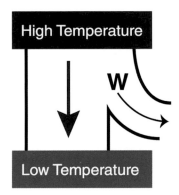

FIGURE 2.17: The energy created when heat flows from hot to cold can be used to do work (W)

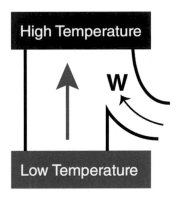

FIGURE 2.18: When energy is added, heat can be transferred from cold to hot

Web Resources

Experiment with pressure, volume, and temperature changes with a piston.
www.mhhe.com/physsci/physical/giambattista/thermo/thermodynamics.html

Physics applets on a variety of science topics, including thermodynamics.
http://jersey.uoregon.edu/

Compare heat conduction through two materials of different thermal conductivities.
http://energy.concord.org/energy2d/thermal-conductivity.html

Explore how energy flows and changes through a system.
http://phet.colorado.edu/en/simulation/energy-forms-and-changes

Learn how the properties of gas (volume, heat, and so on) vary in relation to each other.
http://phet.colorado.edu/en/simulation/gas-properties

A Boyle's law experiment.
www.chm.davidson.edu/vce/gaslaws/boyleslaw.html

A Boyle's law animation.
www.grc.nasa.gov/WWW/k-12/airplane/aboyle.html

A Boyle's law worksheet.
www.grc.nasa.gov/WWW/k-12/BGP/Sheri_Z/boyleslaw_act.htm

A Charles's law experiment.
www.chm.davidson.edu/vce/gaslaws/charleslaw.html

Animated activities and worksheets on gas laws.
www.nclark.net/GasLaws

A Charles and Gay-Lussac's law activity.
www.grc.nasa.gov/WWW/k-12/airplane/glussac.html

A Charles and Gay-Lussac's law animation.
www.grc.nasa.gov/WWW/k-12/airplane/aglussac.html

Relevant Standards

Note: The Next Generation Science Standards *can be viewed online at* www.nextgenscience.org/next-generation-science-standards.

PERFORMANCE EXPECTATIONS

HS-PS3-4

Plan and conduct an investigation to provide evidence that the transfer of thermal energy when two components of different temperature are combined within a closed system results in a more uniform energy distribution among the components in the system (second law of thermodynamics).

SCIENCE AND ENGINEERING PRACTICES

- Plan and conduct an investigation individually and collaboratively to produce data to serve as the basis for evidence, and in the design: decide on types, how much, and accuracy of data needed to produce reliable measurements and consider limitations on the precision of the data (e.g., number of trials, cost, risk, time), and refine the design accordingly.

- Analyze data using tools, technologies, and/or models (e.g., computational, mathematical) in order to make valid and reliable scientific claims or determine an optimal design solution.

- Constructing explanations and designing solutions in 9–12 builds on K–8 experiences and progresses to explanations and designs that are supported by multiple and independent student-generated sources of evidence consistent with scientific ideas, principles, and theories.

- Apply scientific ideas to solve a design problem, taking into account possible unanticipated effects.

CONNECTIONS TO NATURE OF SCIENCE

Science Models, Laws, Mechanisms, and Theories Explain Natural Phenomena

- Theories and laws provide explanations in science.

- Laws are statements or descriptions of the relationships among observable phenomena.

DISCIPLINARY CORE IDEAS

PS1.A: Structure and Properties of Matter

- Structure and Properties of Matter: Matter of any type can be subdivided into particles that are too small to see, but even then the matter still exists and can be detected by other means. A model showing that gases are made from matter particles that are too small to see and are moving freely around in space can explain many observations, including the inflation and shape of a balloon and the effects of air on larger particles or objects.

PS3.A: Definitions of Energy

- The term "heat" as used in everyday language refers both to thermal energy (the motion of atoms or molecules within a substance) and the transfer of that thermal energy from one object to another. In science, heat is used only for this second meaning; it refers to the energy transferred due to the temperature difference between two objects. (secondary to MS-PS1-4)

- The temperature of a system is proportional to the average internal kinetic energy and potential energy per atom or molecule (whichever is the appropriate building block for the system's material). The details of that relationship depend on the type of atom or molecule and the interactions among the atoms in the material. Temperature is not a direct measure of a system's total thermal energy. The total thermal energy (sometimes called the total internal energy) of a system depends jointly on the temperature, the total number of atoms in the system, and the state of the material. (secondary to MS-PS1-4)

CROSSCUTTING CONCEPTS

Patterns

- Different patterns may be observed at each of the scales at which a system is studied and can provide evidence for causality in explanations of phenomena.

Cause and Effect

- Empirical evidence is required to differentiate between cause and correlation and make claims about specific causes and effects.

- Systems can be designed to cause a desired effect.

3

ENERGY

Energy is an essential part of all branches of physics. In mechanics, energy appears as both *potential* and *kinetic energy*. Gravitational potential energy depends on an object's weight and how high the object is raised. An object has kinetic energy if it is moving from one place to another (*translational kinetic energy*) or if it is rotating (*rotational kinetic energy*). When sound is produced, mechanical waves carry energy, while in light waves the photon—a massless particle—carries the energy. With electrical devices, we can change electrical energy to many other forms of energy. Heat energy is involved in changes in temperature and states of matter. The law of conservation of energy tells us that energy does not vanish but only changes form. We encounter this when we speak about energy consumption or energy use.

In the following experiments, we will examine the conservation of energy and how energy can change from one type to another.

Exploration

ENERGY ON WHEELS

The Introductory Energy and Motion Lab, with its car on a track (Figure 3.1), is used to study conservation of mechanical energy. When discussing mechanical energy, it is useful to distinguish between two types of energy: kinetic energy and potential energy.

Procedure

1. Set up the car track so that you can adjust the inclination.

2. Place the speedometer and photogate at the end of the track.

3. Measure the height from which the car starts.

4. Measure the speed of the car with the photogate.

5. Write down the values of the height, h, and the final velocity, v.

6. Repeat the test, but change the car's starting height.

▼
SAFETY NOTE

Wear safety glasses or goggles.

Questions

*   Does the car's starting height affect its final speed? Explain.

*   Draw a graph showing the final speed as a function of starting height.

*   Calculate the square of the final speed values, v^2, and draw a graph of the square of the final speed as a function of starting height.

*   Determine the slope of the curve from the graph. Derive the value of the slope theoretically, assuming that the car's starting potential energy changes completely to kinetic energy.

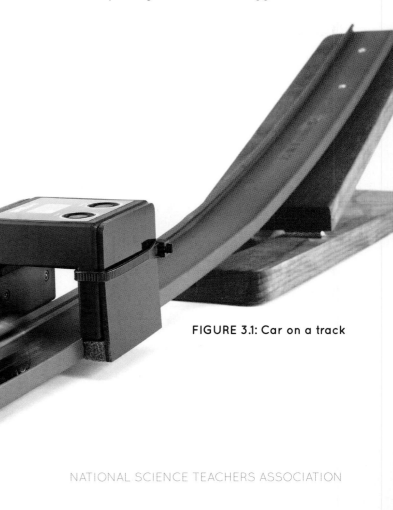

FIGURE 3.1: Car on a track

FIGURE 3.2: Euler's Disc

DANCING DISC

Kinetic energy can be translational, rotational, or both. In this experiment with Euler's Disc (Figure 3.2), we learn what factors affect the magnitude of the rotational energy as well as its conservation.

Procedure

1. Place the mirrored surface securely on a flat surface.

2. Spin the metal disc on the mirrored surface so that it rotates rapidly.

3. Watch the rotation of the disc.

Questions

- Why does the disc spin for so long?

- What determines how much rotational energy there is?

- Does the disc eventually stop? Explain.

▼ SAFETY NOTES

- Wear safety glasses or goggles.

- Use caution when handling the glass and metal mirrors: they may have sharp edges that can cut skin.

HOT SHOT

The physics of collisions is always interesting. In a collision, kinetic energy quickly changes into other forms of energy such as sound or heat. Here we use the Colliding Steel Spheres (Figure 3.3) to learn about this phenomenon.

Procedure

1. Bang the spheres together so that they collide while holding a sheet of paper between them.

2. Repeat the test by changing the force on the spheres and the thickness of the paper.

3. Repeat the test using aluminum foil instead of paper.

Questions

- What happens to the paper where the balls collide? How does the thickness of the paper or the speed of the balls affect the outcome?

- What happens when you use foil instead of paper? How can the pattern in the foil be explained?

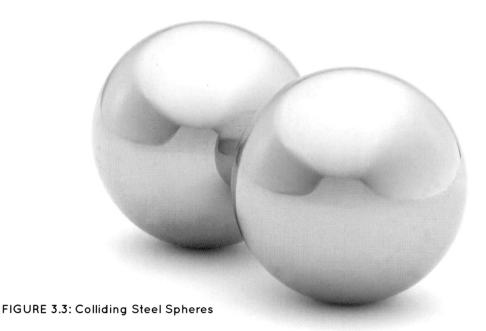

FIGURE 3.3: Colliding Steel Spheres

RADIANT ROTATION

Crookes Radiometer (Figure 3.4) was originally designed to measure radiation pressure. A few years after its invention, it was found that the explanation that had been given for the rotation of the vanes in terms of radiation pressure was not correct. It turned out that the radiometer does not measure radiation pressure, but it does demonstrate some interesting physics.

Procedure

1. Observe the radiometer in direct sunlight outside.

2. Take it inside and warm it up with a hairdryer.

3. Move the radiometer to a place much colder than the room, such as a freezer.

Questions

- What are some possibilities for why the radiometer vanes are rotating?

- Is the direction of rotation in a cold place different from the direction when the radiometer is in the sunlight or heated by the hairdryer?

▼
SAFETY NOTE

Wear safety glasses or goggles.

FIGURE 3.4: Crookes Radiometer

- Wear safety glasses or goggles.

- Use caution when balls are dropped on the floor as they can cause a slip, trip, or fall hazard.

MAGNETIC ACCELERATOR

In nature, the amount of energy in an isolated (or closed) system does not increase or decrease. The energy can only be converted from one form to another. Balls normally lose a lot of kinetic energy during collisions. In the following experiment using the Magnetic Accelerator (Figure 3.5), you will investigate whether that is always the case.

Procedure

1. Set aside the magnetic ball.

2. Put one of the other balls at the top of the track and let it go.

3. When the first ball has settled on the bottom of the track, put the second ball at the top of the track and let it go.

4. Repeat this with the third ball.

5. Repeat this with the magnetic ball that you set aside earlier.

6. Finally, place the fifth ball at the top of the track and let it go.

Questions

- What forms of energy was the potential energy of the first ball converted to?

- How do you explain the results from the last two balls?

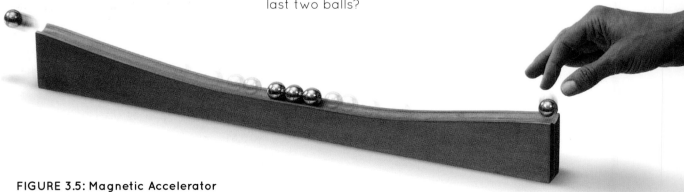

FIGURE 3.5: Magnetic Accelerator

HAPPY / UNHAPPY BALLS

In elastic collisions, the kinetic energy remains unchanged, whereas in inelastic collisions at least some kinetic energy gets transformed to other forms of energy. The properties of the materials that the colliding bodies are made of determine whether the collision is more elastic or inelastic. The Happy / Unhappy Balls (Figure 3.6) are used to demonstrate this.

Procedure

1. Drop the two balls onto the floor and see what happens.

2. Simultaneously roll the balls on the floor or down an incline and see what happens.

3. Place the balls in a freezer for at least a few hours and perform the drop and rolling tests again.

Questions

- Describe the behavior of each ball when you bounce it. Why does the unhappy ball act differently than the happy ball? Where does the energy of the unhappy ball go?

- Describe the behavior of each ball when you roll it. Why does the unhappy ball act differently than the happy ball?

- Does the same explanation work for both the balls' bouncing behavior and their rolling behavior?

- Did the results change when the balls were frozen? Explain.

▼
SAFETY NOTES

- Wear safety glasses or goggles.

- Use caution when balls are dropped on the floor as they can cause a slip, trip, or fall hazard.

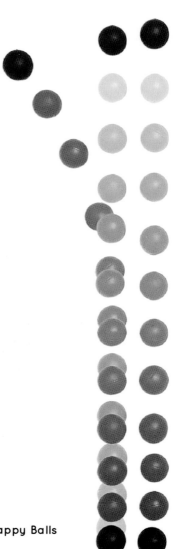

FIGURE 3.6: Happy / Unhappy Balls

FIGURE 3.7:
Dropper Popper

DROPPER POPPER

The Dropper Popper (Figure 3.7) works somewhat
like a super ball. It is dropped onto the floor,
and it bounces back from the floor.

Procedure

1. Load the Popper by turning its edges down.

2. Drop the Popper from a height of 1 m.

Questions

- How high does the Popper bounce relative to its
  starting height? Why?

- What are the Popper's sources of energy?

ASTROBLASTER

The Astroblaster (Figure 3.8) is a toy consisting of five bouncy balls of different sizes stacked on a stick.

Procedure

Drop the Astroblaster from a height of about 0.5 m.

Questions

- One of the balls acts differently from the rest. Explain what it does and why.

- Explore what percentage of the initial energy the ball can end up having. (A video camera might help you with this investigation.)

FIGURE 3.8: Astroblaster

▼
SAFETY NOTE

Wear safety glasses or goggles.

Analysis

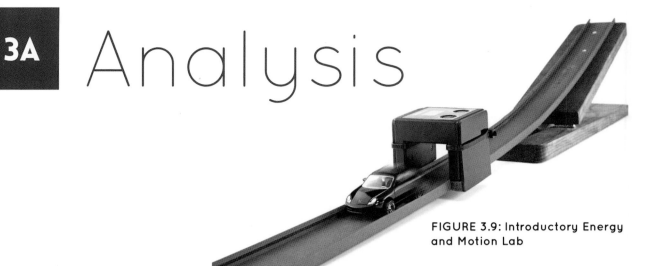

FIGURE 3.9: Introductory Energy and Motion Lab

ENERGY ON WHEELS

The car in the Introductory Energy and Motion Lab (Figure 3.9) demonstrates how potential energy becomes partly transformed to kinetic energy. You perform work when you lift an object (in this case, the car) because you are working against the force of gravity. The work (W) done by a constant force, F, is equal to the force times the distance, x, over which the force is applied:

$$W = Fx$$

As the object is raised, the energy from the work done is stored as additional potential energy, ΔE_p.

$$\Delta E_p = W$$

The average lifting force equals the weight of the object, mg.

$$F = mg$$

When lifting an object, the distance, x, equals the height, h, so, combining these equations, we have that the change in potential energy is

$$\Delta E_p = mgh.$$

When the car rolls down the track, potential energy is converted to kinetic energy. A greater starting height means that the car has more potential energy, so when the car gets to the bottom of the track, more energy has been transformed to kinetic energy. Therefore, increasing the starting height will increase the final speed of the car.

In reality, not all of the potential energy gets converted into kinetic energy. Some of the energy is lost due to friction, air resistance, sound, and heat. However, these losses are fairly small and, because the length of the track doesn't change, these losses are approximately constant as the starting height is changed.

The resultant kinetic energy is equal to the potential energy:

$$mgh = \tfrac{1}{2}mv^2$$

From this, the lifting height can be related to the final velocity:

$$h = \frac{v^2}{2g}$$

or

$$v^2 = 2gh.$$

When the square of the final velocity is graphed as a function of the starting height, a straight line is formed with a slope of $2g$.

DANCING DISC

Euler's Disc (Figure 3.10) is edge-rounded and polished so that the frictional force is minimized during its rotation on the mirrored surface. The rotation continues not only because there is little friction but also because the metal disc has a large mass. A large mass increases the moment of inertia and therefore the rotational energy. The rotational energy, E_{rot}, is related to the moment of inertia, I, and the angular velocity, ω, as follows:

$$E_{rot} = \frac{1}{2}I\omega^2$$

Finally, the rotation of the metal disc eventually stops because of work done by friction and air resistance. Some of the energy gets converted to sound and some to heat.

Extension

If you have the toys that go with Chapter 8 ("Angular Momentum"), then you have a plastic stick called a rattleback or celt (it is flat on the top and curved on the bottom). Spin the rattleback first clockwise and then counterclockwise. What happens?

In the rattleback, the center of mass does not exactly coincide with the rotational axis, and the point at which the rattleback touches the table is not located on the rotational axis. This means that the friction at this point produces a torque. This torque slows the rotation, makes the rattleback wobble, and then causes the rattleback to turn in the opposite direction. The wobbling movement changes the rotational motion because of the shape of the rattleback, which produces a torque that makes the rattleback twist a little bit as it wobbles, and this twisting makes it rotate more. Watch it closely to see the motion change among all three axes.

FIGURE 3.10: Euler's Disc

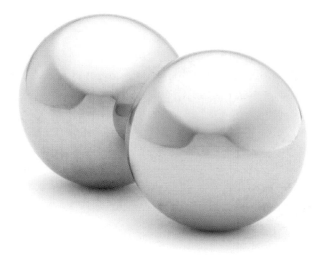

HOT SHOT

In the "Hot Shot" exploration, students watch as two metal spheres collide (Figure 3.11). The impact of steel spheres is very close to a perfectly elastic collision, meaning very little of the balls' kinetic energy is lost. Some of the kinetic energy, however, is converted to other forms of energy such as heat and sound. The spheres are massive (weighing about 1 lb each), and therefore the kinetic energy in the collision is rather high. The spherical shape of the steel spheres and their small radii causes the energy to be concentrated in a very small volume. Heat energy in the collision burns the paper or melts the aluminum foil.

Hitting the spheres together with more force increases the amount of heat energy. When the collision energy increases, a larger hole is burned in the paper or the pattern of waves in the aluminum is larger. In that case, the greater amount of energy melted more of the aluminum and made a larger wave pattern. The heat is quickly transmitted to the environment as the aluminum solidifies.

FIGURE 3.11: Colliding Steel
Spheres

RADIANT ROTATION

Crookes radiometer (Figure 3.12; also known as a *light mill*) is a good example of the conversion of heat energy to kinetic energy. In this case, the explanation is not quite so simple. The radiometer contains a partial vacuum, with an air pressure of perhaps only 10^{-5} that of normal atmospheric pressure. The vanes in a radiometer are white on one side and black on the other. The black surfaces absorb radiation (e.g., either visible light, such as sunlight, or infrared) better and warm up faster than the white surfaces.

One of the earliest explanations was that, because of the warmth of the black surface, the air molecules near that surface heat up and begin to move faster. Some of the molecules then bounce away from the surface, causing an impulse that pushes the vane. A sufficient number of air molecules bouncing off is what causes the vanes of the radiometer to rotate in the direction of the white surfaces.

The problem with this explanation is that it turns out that even though the faster-moving air molecules on the black side create more force, those fast-moving molecules also do a better job of keeping other air molecules from hitting the black surface. The net effect is that there should be no extra force pushing on the black side. But it turns out that there *is* an excess force on the black side at the *edges* of the vanes. That's actually where the force arises that turns the vanes. (One can't deduce this just by watching the radiometer. Here you are mainly looking for plausible explanations that have some supporting evidence based on the observations and the students' prior knowledge.)

When the radiometer is cooled, the black surfaces emit heat more efficiently and therefore become cooler than the white surfaces. Then, with the edge effects working in the opposite direction, the direction of the rotation is opposite, so that the black sides are the leading sides.

HAPPY / UNHAPPY BALLS

The two balls in this exploration look similar (Figure 3.14). However, the balls are made of different materials. The difference is detected when the balls are dropped or rolled on the floor. The "happy" ball bounces, but the "unhappy" ball does not bounce. In addition, the happy ball rolls more easily than the unhappy ball. Conversely, if the balls are frozen, the unhappy ball bounces higher than the happy ball.

Rubber is characterized by the fact that it rapidly returns to its original shape after it is deformed.

FIGURE 3.14:
Happy / Unhappy Balls

Deformation occurs when the ball strikes the ground. The happy ball is made of a kind of rubber that returns to its original form much faster than the rubber of the unhappy ball. The kinetic energy that the balls have before a collision is converted to heat (due to friction causing random motion of the molecular structures), and also to potential energy of molecular structure (like compressing a spring), and back again to kinetic energy. With the happy ball, more potential energy is converted to kinetic energy. With the unhappy ball, a lot of the energy is converted to heat because of internal friction.

The balls' different responses are related to the level of elasticity in the collision. The happy ball bounces because the collision is elastic. The unhappy ball does not bounce; the collision is inelastic. When the balls roll on the floor, the unhappy ball rolls more slowly than the happy ball because there is more friction between the unhappy ball and the floor. The properties of the different rubber compounds change as the balls are frozen. The happy ball loses its elasticity, whereas the unhappy ball's elasticity is increased.

FIGURE 3.13: Magnetic Accelerator

MAGNETIC ACCELERATOR

The Magnetic Accelerator (Figure 3.13) activity explores the impact magnets have on potential energy. Each ball's potential energy is converted into kinetic energy and then into sound energy and, due to resistance forces such as friction and air resistance, into heat. The fourth ball, which is magnetic, accelerates just before its impact with other balls, thereby producing enough energy to cause the end ball to fly off the end of the ramp. The same occurs with the last ball, for which the magnetic ball is at rest, because the approaching ball accelerates rapidly because of the magnetic force pulling it forward.

The collisions are almost elastic, and the kinetic energy of the ball is transferred to the last ball in line, which is far enough from the magnetized ball to not be noticeably affected by its magnetic field. The last ball has so much energy from the collision that it easily leaves the ramp.

FIGURE 3.16:
Astroblaster

FIGURE 3.15: Dropper Popper

DROPPER POPPER

At the start of the experiment, the edges of the Dropper Popper (Figure 3.15) are turned down. This takes energy, and that energy is stored in the Popper. When the Popper hits the ground, the stored (potential) energy is released as the edges return to their original position. This means that not only is the gravitational potential energy converted to kinetic energy as the Popper falls, but that additional kinetic energy is also obtained from the stored potential energy in the flexed edges of the Popper. Consequently, the Popper jumps to a height greater than its original height.

ASTROBLASTER

The balls of the Astroblaster (Figure 3.16), which are stacked on a stick, are very elastic and bouncy, making the collisions between them almost elastic. When the Astroblaster hits the floor, kinetic energy is transferred from the lower, more massive balls, up to the higher, smaller balls. If the kinetic energy is transferred from a more massive ball to a less massive ball, the less massive ball must move faster. In other words, if $\frac{1}{2}mv^2$ is constant, then the ball with the lower mass must have the higher velocity. The ball that bounced highest did so for that reason and because it ended up with much of the kinetic energy.

Another experiment you can perform with the Astroblaster is to measure how high the stack of balls bounces with and without the top ball. This will tell you something about the amount of energy taken from the other balls, since they won't bounce as high when the top ball is included.

Web Resources

Some basic concepts and terminology of kinetic energy.
www.physicsclassroom.com/Class/energy/U5L1c.cfm

Learn about kinetic energy, potential energy and friction by creating simulations of skateboarding ramps.
http://phet.colorado.edu/en/simulation/energy-skate-park-basics

See how a pendulum gains and loses kinetic and potential energy as it swings.
www.clrn.org/weblinks/details.cfm?id=6646

Explore elastic collisions of balls in two dimensions. Change the mass ratio and the impact parameter.
http://physics.usask.ca/~pywell/p121/Notes/collision/collision.html

Investigate the translation and rotation of a rolling disk.
www.schulphysik.de/suren/Applets/Kinematics/Roll/RollApplet.html

See the position, velocity, and acceleration vectors for a point on the rim of a rolling object.
www.phys.hawaii.edu/~teb/java/ntnujava/FreeRolling/FreeRolling.html

Relevant Standards

Note: The Next Generation Science Standards *can be viewed online at* www.nextgenscience.org/next-generation-science-standards.

PERFORMANCE EXPECTATIONS

HS-PS3-1

Create a computational model to calculate the change in the energy of one component in a system when the change in energy of the other component(s) and energy flows in and out of the system are known.

HS-PS3-2

Develop and use models to illustrate that energy at the macroscopic scale can be accounted for as a combination of energy associated with the motions of particles (objects) and energy associated with the relative positions of particles (objects).

HS-PS3-3

Design, build, and refine a device that works within given constraints to convert one form of energy into another form of energy. [Integrates science and engineering]

HS-PS3-5

Develop and use a model of two objects interacting through electric or magnetic fields to illustrate the forces between objects and the changes in energy of the objects due to the interaction.

SCIENCE AND ENGINEERING PRACTICES

Developing and Using Models

Modeling in 9–12 builds on K–8 and progresses to using, synthesizing, and developing models to predict and show relationships among variables between systems and their components in the natural and designed worlds.

- Develop and use a model based on evidence to illustrate the relationships between systems or between components of a system.

Planning and Carrying Out Investigations

Planning and carrying out investigations to answer questions or test solutions to problems in 9–12 builds on K–8 experiences and progresses to include investigations that provide evidence for and test conceptual, mathematical, physical, and empirical models.

- Plan and conduct an investigation individually and collaboratively to produce data to serve as the basis for evidence, and in the design: decide on types, how much, and accuracy of data needed to produce reliable measurements and consider limitations on the precision of the data (e.g., number of trials, cost, risk, time), and refine the design accordingly.

Using Mathematics and Computational Thinking

Mathematical and computational thinking at the 9–12 level builds on K–8 and progresses to using algebraic thinking and analysis, a range of linear and nonlinear functions including trigonometric functions, exponentials and logarithms, and computational tools for statistical analysis to analyze, represent, and model data. Simple computational simulations are created and used based on mathematical models of basic assumptions.

- Create a computational model or simulation of a phenomenon, designed device, process, or system.

Constructing Explanations and Designing Solutions

Constructing explanations and designing solutions in 9–12 builds on K–8 experiences and progresses to explanations and designs that are supported by multiple and independent student-generated sources of evidence consistent with scientific ideas, principles, and theories.

Design, evaluate, and/or refine a solution to a complex real-world problem, based on scientific knowledge, student-generated sources of evidence, prioritized criteria, and tradeoff considerations.

CONNECTIONS TO NATURE OF SCIENCE

Science Models, Laws, Mechanisms, and Theories Explain Natural Phenomena

- Theories and laws provide explanations in science.

- Laws are statements or descriptions of the relationships among observable phenomena.

DISCIPLINARY CORE IDEAS

PS3.A: Definitions of Energy

- Energy is a quantitative property of a system that depends on the motion and interactions of matter and radiation within that system. That there is a single quantity called energy is due to the fact that a system's total energy is conserved, even as, within the system, energy is continually transferred from one object to another and between its various possible forms.

- At the macroscopic scale, energy manifests itself in multiple ways, such as in motion, sound, light, and thermal energy.

- These relationships are better understood at the microscopic scale, at which all of the different manifestations of energy can be modeled as a combination of energy associated with the motion of particles and energy associated with the configuration (relative position of the particles). In some cases the relative position energy can be thought of as stored in fields (which mediate interactions between particles). This last concept includes radiation, a phenomenon in which energy stored in fields moves across space.

PS3.B: Conservation of Energy and Energy Transfer

- Conservation of energy means that the total change of energy in any system is always equal to the total energy transferred into or out of the system.

- Energy cannot be created or destroyed, but it can be transported from one place to another and transferred between systems.

- Mathematical expressions, which quantify how the stored energy in a system depends on its configuration (e.g., relative positions of charged particles, compression of a spring) and how kinetic energy depends on mass and speed, allow the concept of conservation of energy to be used to predict and describe system behavior.

- The availability of energy limits what can occur in any system.

- Uncontrolled systems always evolve toward more stable states—that is, toward more uniform energy distribution (e.g., water flows downhill, objects hotter than their surrounding environment cool down).

PS3.C: Relationship Between Energy and Forces

- When two objects interacting through a field change relative position, the energy stored in the field is changed.

PS3.D: Energy in Chemical Processes

- Although energy cannot be destroyed, it can be converted to less useful forms—for example, to thermal energy in the surrounding environment.

CROSSCUTTING CONCEPTS

Energy and Matter

- Changes of energy and matter in a system can be described in terms of energy and matter flows into, out of, and within that system.

- Energy cannot be created or destroyed—only moves between one place and another place, between objects and/or fields, or between systems.

VISIBLE LIGHT AND COLORS

Color theory is an area of physics that can explain how the appearance of colors changes when colors are mixed, for example on a computer screen or with a printing press. Visible light can be divided into different ranges of wavelengths, which we see as different colors. The color of an object is usually the color of the light that the surface of the object is reflecting when viewed in white light. Some of the white light is absorbed and some is reflected from the surface—in a particular direction if the surface is shiny or scattered in many directions if the surface is rough.

In the eye, there are light-sensitive cells called *rods* and *cones*. The rod cells can detect the brightness of the light but do not detect color; the cone cells detect the color of the light. There are three kinds of cone cells, and each is sensitive to different ranges of wavelengths. One type of cone cell is sensitive to short wavelengths (such as blue light), one type to medium wavelengths (such as green light), and one type to long wavelengths (such as red light). The sensitivity of the rod and cone cells will vary in different situations. The brain interprets the signals from the rods and cones and creates the sensation of different colors.

Although sunlight is usually seen as white, it actually consists of many different colors. You can see this by using a prism or a spectroscope. In a prism, the various wavelengths refract by different amounts and create a spectrum with all the colors of the rainbow from violet to red.

In this chapter, we will study the formation and sensation of color. We will also investigate primary, secondary, and complementary colors.

FIGURE 4.1:
Quantitative Spectroscope

SPECTROSCOPE

With the Quantitative Spectroscope (Figure 4.1) you can study the spectra of different light sources and the wavelengths of various colors.

Procedure

1. Point the spectroscope toward sunlight or an incandescent lamp.

2. Read the wavelengths of the colors from the scale in the spectroscope.

3. Check the spectrum of a fluorescent lamp or an energy-saving lamp.

Questions

• What colors can you see in the spectrum?

• What are the differences between the spectrum of a lightbulb and that of a fluorescent lamp?

• How can these differences be explained?

FIGURE 4.2: Colored shadows from Primary
Color Light Sticks and color filters

ADDING AND SUBTRACTING COLORS

**With flashlights and color filters,
we can study the formation of colors
and colored shadows (Figure 4.2).**

Procedure

1. Build three different-color light sources
   by attaching the RGB filters to the front
   of three flashlights.

2. Dim the lights in the classroom.

3. Point the red and blue lights toward a
   whiteboard or screen so that the colors
   overlap.

4. Repeat with other color combinations
   (red and green, blue and green).

5. Point all three colors at the screen so
   that they overlap.

6. Put an object in front of the red light.
   Look at the color of the shadow.

7. Repeat with the blue light and then the
   green light.

Questions

- What are the three primary colors of
  light?

- You may have previously seen the term
  RGB. Where?

- Make a diagram of the colors you have
  seen in your work here. All the primary
  colors and their combinations should
  be included.

SAFETY NOTE

Use caution when moving
around the lab with dimmed
lights. Make sure there are
no trip, slip, or fall hazards
on the floor.

FIGURE 4.3: RGB Snap Lights and Spinner

FIGURE 4.4: Tape placement for the three colored tubes

▼
SAFETY NOTES

- Wear safety goggles.

- Use caution when moving around the lab with dimmed lights. Make sure there are no trip, slip, or fall hazards on the floor.

COMBINING COLORS

Next, we will use the RGB Snap Lights and Spinner (Figure 4.3) to look at the three primary colors and the formation of white light. We will also discuss color subtraction.

Procedure

1. Bend the snap-light tubes (red, green, and blue) so that the thin capsules break inside the tube.

2. Attach the snap lights to the spinner and dim the lights in the classroom.

3. Spin the device fast and see what color appears.

4. Take a piece of black tape, and put it around the end of the red tube. Spin the spinner again and notice the difference in color.

5. Then put a piece of tape close to the end of the green tube, but not as far as the tape on the red tube (Figure 4.4). Spin the spinner and observe the new color.

6. Once again, take some tape and cover a small area on the blue-colored tube as shown in Figure 4.4.

Note: The RGB Snap Lights can be used repeatedly for up to 24 hours if they are stored in a cold place between uses.

Questions

- What color was observed when spinning the first time? Why?

- What is the color on the edge of the spinning circle when there is black tape on the red tube?

- What about when the green tube is covered?

- What about when the blue tube is covered?

- What do you think the letters CMYK stand for?

SPECTROSCOPE

Spectroscopy is an area of physics that analyzes radiation—both visible and invisible kinds of radiation. Here you are examining visible light. In a spectroscope (Figure 4.5), light passes through a narrow slit to a diffraction grating. A diffraction grating has many closely spaced grooves through which light diffracts, deflecting different wavelengths at different angles to separate the various colors or wavelengths. The colors of light are directed toward the scale of the spectroscope so that you can read off the wavelengths.

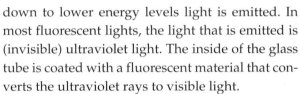

FIGURE 4.5: Quantitative Spectroscope

In the first step of the exploration, the spectroscope is used to examine natural or incandescent light. When analyzing the radiation emitted by an incandescent light (in which light is emitted by a hot filament), a continuous spectrum is observed. This spectrum shows all of the wavelengths of visible light from violet to red.

In the last step, you looked at a different type of light. Fluorescent lights, light-emitting diodes, and many energy-saving lights produce a spectrum that is not continuous but shows only specific lines or bands of color. The spectral lines produced by a fluorescent light can be explained by the formation of excited states of atoms or molecules. Electrons are temporarily transferred to higher energy levels, and when they jump back down to lower energy levels light is emitted. In most fluorescent lights, the light that is emitted is (invisible) ultraviolet light. The inside of the glass tube is coated with a fluorescent material that converts the ultraviolet rays to visible light.

All the chemical elements have a unique set of spectral lines, so the spectrum can be used to identify the elements. A mostly dark spectrum with some bright, colored lines is called an *emission spectrum*. A different type of experiment (not demonstrated here) can show a mostly continuous spectrum containing dark lines, which is called an *absorption spectrum*. The dark lines are called *absorption lines*. This results from light passing through a substance and that substance absorbing certain wavelengths (or colors) of light.

ADDING AND SUBTRACTING COLORS

In this activity, you used flashlights and color filters to produce colored shadows (Figure 4.6). The filters came in the three *primary colors*, which cannot be formed by mixing other colors. For colors of light, these primary colors (red, blue, and green) can be combined according to the rules of color addition.

FIGURE 4.6: Colored shadows

First, you combined the three primary colors in pairs to obtain the *subtractive primary colors*: cyan, magenta, and yellow. A combination of equal amounts of red and green light appears yellow (to people with normal color vision) even though it doesn't contain any actual yellow wavelengths of light. Similarly, magenta (formed by combining red and blue light) does not contain any magenta light wavelengths, and cyan (formed by combining green and blue light) does not contain any cyan light wavelengths. When all three colors are combined, we get white light (Figure 4.7).

In step 6 of this experiment, you observed the colors of shadows. When one of the primary colors was blocked, you saw the differently colored shadows as yellow (when blue was blocked), cyan

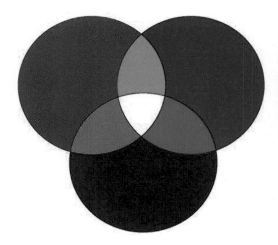

FIGURE 4.7: Combining lights in the three primary colors produces white light.

(when red was blocked), and magenta (when green was blocked).

You may recognize the term RGB, which was used in the exploration. Colors displayed on a computer monitor or TV screen are created using the RGB method. Each screen pixel is a single color of light—red, green, or blue. The colors shown on the screen are formed by combinations of the three primary colors displayed with the appropriate intensities.

Another color term you might know is CMYK. The subtractive colors (which you can obtain by subtracting one of the primary colors from white light as mentioned earlier) are cyan (C), magenta (M), and yellow (Y). If you subtract all three primary colors, then you have black (K). (Black is symbolized by K, which stands for "key," because in the printing industry the black plate was used as a reference to key, or align, the other color plates.) So the CMYK color system is called a subtractive color system not merely because you can obtain these colors by removing a primary color from white light, but also because when you combine these subtractive colors, for example, when mixing paints, additional wavelengths (or colors) are removed from the reflected light. For example, if you combine yellow paint (which usually reflects not only yellow light but also some red and green light) with blue or cyan paint (which absorbs red light) the light color that gets reflected the most will be green.

Subtractive color filters remove some part of the white light spectrum. A yellow filter blocks the passage of the blue light, a magenta filter blocks green, and a cyan filter blocks red. Black is used for the parts of an image that contain no light (Figure 4.8).

COMBINING COLORS

Putting tape over one of the colors in the RGB Snap Light and Spinner (Figure 4.9) results in your seeing the subtractive colors cyan, magenta, and yellow (CMY).

Colors used by printers are usually abbreviated by CMYK, in which K stand for "key." Four-color printing means, therefore, that the four colors of ink or dye are used for the formation of color images. When using CMYK colors, say in image-editing software, the intensity of the four CMYK colors is saved for each pixel.

**FIGURE 4.9:
RGB Snap Lights
and Spinner**

Finally, a word about the color indigo. Most people have heard of indigo from the well-known mnemonic "ROY G BIV," which describes seven colors identified by Isaac Newton (Figure 4.10). However, those names are arbitrary, and nowadays you almost never hear anyone use the word indigo except in connection with "ROY G BIV." Actually there's really no good reason to include indigo as a major color. It has no physical significance as a primary or secondary color. Now many teachers use the mnemonic "ROY G BV" (without the I) instead.

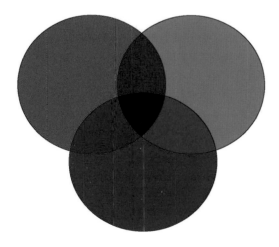

FIGURE 4.8: Subtractive color filters combine in pairs to produce other colors or, with all three filters, no color at all (i.e., black).

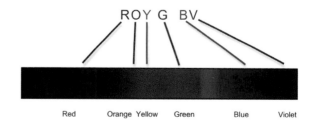

FIGURE 4.10: The primary and secondary colors

Using Mathematics and Computational Thinking

Mathematical and computational thinking at the 9–12 level builds on K–8 and progresses to using algebraic thinking and analysis, a range of linear and nonlinear functions including trigonometric functions, exponentials and logarithms, and computational tools for statistical analysis to analyze, represent, and model data. Simple computational simulations are created and used based on mathematical models of basic assumptions.

- Create a computational model or simulation of a phenomenon, designed device, process, or system.

Constructing Explanations and Designing Solutions

Constructing explanations and designing solutions in 9–12 builds on K–8 experiences and progresses to explanations and designs that are supported by multiple and independent student-generated sources of evidence consistent with scientific ideas, principles, and theories.

- Design, evaluate, and/or refine a solution to a complex real-world problem, based on scientific knowledge, student-generated sources of evidence, prioritized criteria, and tradeoff considerations.

CONNECTIONS TO NATURE OF SCIENCE

Science Models, Laws, Mechanisms, and Theories Explain Natural Phenomena

- Theories and laws provide explanations in science.

- Laws are statements or descriptions of the relationships among observable phenomena.

DISCIPLINARY CORE IDEAS

PS4.B: Electromagnetic Radiation

- Atoms of each element emit and absorb characteristic frequencies of light. These characteristics allow identification of the presence of an element, even in microscopic quantities.

PS4.B: Electromagnetic Radiation

- Electromagnetic radiation (e.g., radio, microwaves, light) can be modeled as a wave of changing electric and magnetic fields or as particles called photons. The wave model is useful for explaining many features of electromagnetic radiation, and the particle model explains other features.

- When light or longer wavelength electromagnetic radiation is absorbed in matter, it is generally converted into thermal energy (heat). Shorter wavelength electromagnetic radiation (ultraviolet, X-rays, gamma rays) can ionize atoms and cause damage to living cells.

- Photoelectric materials emit electrons when they absorb light of a high-enough frequency.

PS4.C: Information Technologies and Instrumentation

- Multiple technologies based on the understanding of waves and their interactions with matter are part of everyday experiences in the modern world (e.g., medical imaging, communications, scanners) and in scientific research. They are essential tools for producing, transmitting, and capturing signals and for storing and interpreting the information contained in them.

ESS1.A: The Universe and Its Stars

- The study of stars' light spectra and brightness is used to identify compositional elements of stars, their movements, and their distances from Earth.

CROSSCUTTING CONCEPTS

Energy and Matter

- Changes of energy and matter in a system can be described in terms of energy and matter flows into, out of, and within that system.

- Energy cannot be created or destroyed—only moves between one place and another place, between objects and/or fields, or between systems.

RESONANCE

Resonance—a system's tendency to oscillate at greater amplitudes at certain frequencies—is a common phenomenon in physics. Any system that can oscillate (e.g., a pendulum) will most easily do so at the system's natural frequency. External forces can cause the system to oscillate at any frequency, but if an external force acts on the system at its natural frequency, the amplitude of the oscillations will grow. The system is then in resonance with the external force. Some systems can have many different resonant frequencies. For a vibrating string or sound resonating in a tube, the lowest resonant frequency is called the *fundamental frequency*, or the *first harmonic*. Higher resonant frequencies would include the *second harmonic*, *third harmonic*, and so on.

▼
SAFETY NOTE

Wear safety glasses or goggles.

FIGURE 5.1: Standing Wave Apparatus

STANDING WAVE

In musical instruments, resonance occurs and a sound is created as a result of what is called a standing wave. Here you'll learn something about standing waves using the Standing Wave Apparatus (Figure 5.1).

Procedure

1. Install a AA battery in the Standing Wave Apparatus.

2. Hold the string from the free end.

3. Experiment with holding the string in different places.

4. See how nodes and antinodes are formed in the string.

Questions

- How many nodes and how many antinodes are formed in the string?

- How can you get the basic fundamental wave (having only one antinode) to form in the string?

- Explain how a standing wave is formed in a string.

BOOMWHACKERS

Boomwhackers (Figure 5.2) let you explore open and closed pipe resonances. In this experiment you will investigate what influence the length of a pipe has on the natural frequency of the standing wave and the musical pitch that you hear. Boomwhackers will help you understand how a standing wave is formed in open and closed pipes.

Procedure

1. Explore the different notes you can create with Boomwhackers.

2. How does the length of the pipe affect what note (pitch) is produced?

3. Place the cap on the pipe and investigate how it affects the pitch.

Questions

- How does the length of the pipe affect the pitch?

- How are the fundamental frequency and second harmonic frequency created in a pipe?

- What difference does the cap make to the pitch? Why?

- Make a drawing of how standing waves are created in each situation.

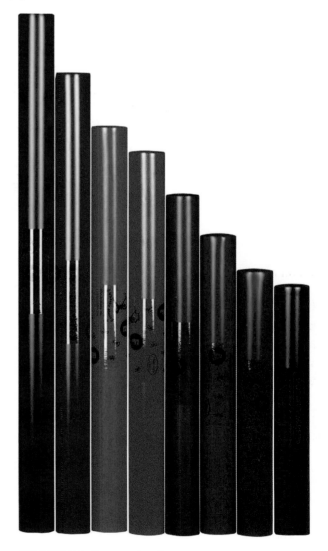

FIGURE 5.2: Boomwhackers

SINGING RODS

A resonance can create a significant swaying motion in mechanical systems. The amplitude of the vibrations can grow when something vibrates with the system's natural frequency, and if the amplitude is high enough, the system can even break. In this experiment, a loud sound is created with the Singing Rods (Figure 5.3) as a result of resonance.

Procedure

1. Hold the rod between your thumb and forefinger at the marked node.

2. Place a little bit of rosin (included) on the fingers of the other hand.

3. Rub the rod with your rosined fingers from the middle to the end. Pinch the rod between a finger and thumb with rosin powder on them. Squeeze with light to moderate pressure and pull toward the end of the rod. Experiment with different amounts of pinching pressure until you get a loud sound.

4. Try it again while holding the rod in different places.

Questions

- What makes the rod sing?

- How does the place where you hold the rod change the pitch and loudness of the sound?

- Measure the frequency of the sound and calculate the speed of sound in the rod.

FIGURE 5.3: Singing Rod

MUSIC BOX

With the Music Box (Figure 5.4), acoustic resonance and transfer of energy (e.g., sound waves) can be explored. Sometimes a musical instrument has more than one vibrating part. For example, with an acoustic guitar the strings vibrate and so does the wooden body of the guitar. Which part does most of the sound come from?

▼
SAFETY NOTE

Make sure there is nothing breakable in the vicinity.

Procedure

1. Wind up the music box, and listen to how it sounds.

2. Place the music box against a desk and listen again.

3. Place the music box against other surfaces and see what difference each makes in the loudness of the sound.

4. Place the music box against your elbow and gently put your finger in your ear. (Use a finger on the hand of the same arm as the elbow that the music box is pressed against.)

FIGURE 5.4: Music Box

Question

•   Why does the volume (i.e., loudness) of the music box vary with different surfaces?

FIGURE 5.5: Sound Pipe

SOUND PIPE

The Sound Pipe (Figure 5.5) can be used to demonstrate the formation of sound, standing wave motion, and resonance. When you twirl the sound pipe, different standing waves are generated with resonant pitches. The frequency of these sounds depends on how long the pipe is and how fast the other end of the pipe moves.

Procedure

1. Vary the speed of twirling and listen to how the speed affects the frequency (pitch) of the sound.

2. Holding one end of the tube, spin the tube above your head.

3. Spin slowly at first, then faster.

4. Inflate a plastic bag and attach it to the lower end of the sound pipe.

5. Twirl the pipe around again.

Questions

- Which way is the air moving in the pipe when you twirl it? Why?

- What creates sound in the pipe?

- What happens to the frequency of the sound when you twirl the pipe faster?

▼
SAFETY NOTES

- Wear safety glasses or goggles.

- Make sure the trajectory path of the Sound Pipe is clear of observers, fragile equipment, and so on, to prevent damage or injury.

Analysis

STANDING WAVE

A standing wave (Figure 5.6) is formed when two waves that are moving in opposite directions collide and interfere. In musical instruments, there are parts that vibrate, such as violin strings, or parts in which the air vibrates, such as the pipes in a pipe organ. There are natural frequencies at which these vibrations occur. When a wave traveling through some medium, such as air, is reflected at the end or boundary and collides with a similar wave (coming from the same source), an interference wave is formed. There are certain places on the string or in the air where no motion is occurring. Such places are called *nodes*. Places where the string or air is moving the most, usually midway between the nodes, are called *antinodes*. In the interference wave, the nodes and antinodes remain at the same positions, so it looks like the wave is not moving. This wave is called a standing wave.

In the exploration, an off-center electric motor creates a vibration which, in turn, creates a transverse wave on the string. A transverse wave has oscillations perpendicular to the direction of propagation. In this case, the direction of propagation is from the motor up to the place where the string is suspended (i.e., your hand), and then the wave reflects back down to the motor. The reflected wave and the wave from the motor interfere with each other. Where the opposing waves are always completely out of phase they create nodes, and where they are in phase, they reinforce each other, creating antinodes.

Even though the standing wave appears not to be moving, it can still transfer energy. The standing wave in a guitar string makes the air close to the string move, and it makes the body of the guitar vibrate. These vibrations of the wood make the

FIGURE 5.6: Standing Wave Apparatus

air near the guitar vibrate, transferring energy to the surroundings via sound waves.

Standing waves play a significant role in musical instruments when sound waves are formed by standing waves on strings (e.g., violin, guitar) and in wind instruments (e.g., flute). The guitar and violin strings are attached at each end, where there is a node, and a standing wave is formed between these two end nodes. In wind instruments there is an antinode at the instrument's open end(s).

BOOMWHACKERS

Boomwhackers (Figure 5.7) are tubes that, with no caps on the ends, have both ends open. The air is free to move at the open ends so, as a standing wave is created, there is an antinode at each end. A standing wave with the fundamental wavelength (also known as the first, or $N = 1$, harmonic) has a wavelength equal to twice the length of the tube. The wavelength, λ, of a sinusoidal waveform traveling with constant speed is given by $\lambda = v \, / \, f$, where v is the speed of sound and f is the frequency.

The speed of sound is constant in the classroom, so only the length of the Boomwhacker tube affects the frequency of the sound. A standing wave in a longer tube has a longer wavelength and therefore creates sound with a lower frequency and lower pitch. Therefore, the longest tube has the lowest pitch.

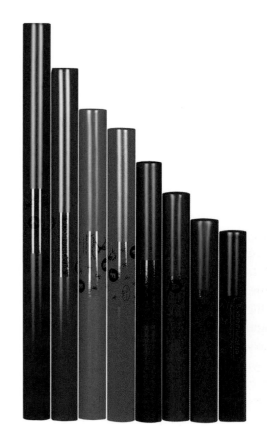

FIGURE 5.7: Boomwhackers

The second ($N = 2$) harmonic is created when the standing wave has two nodes in the tube (as shown in Figure 5.8). Then the wavelength of the sound is the same as the length of the tube. The frequency is therefore twice the fundamental frequency, and the pitch is higher than the fundamental. A higher pitch can be achieved either by producing a higher harmonic or by using a shorter tube.

For the last test, a cap is placed on the end of the Boomwhacker. This has the effect of doubling the wavelength of the standing wave, and therefore lowering (halving) the frequency, so the pitch is lowered. See the Figure 5.8 for examples of standing waves in open and closed tubes.

Standing waves with wavelength λ in tubes of length L

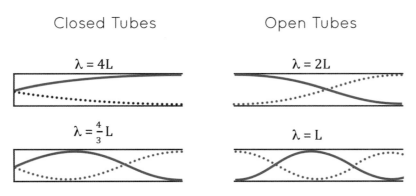

Closed Tubes

$\lambda = 4L$

$\lambda = \frac{4}{3}L$

Open Tubes

$\lambda = 2L$

$\lambda = L$

FIGURE 5.8: Standing waves in open and closed tubes.

Note: This diagram appears to show transverse waves, but the sound waves in the tubes are actually compressional (or longitudinal) waves. Think of these curves as showing how much the air molecules move in different parts of the tube.

SINGING RODS

When you rub the Singing Rod (Figure 5.9) with your resined fingers, the friction between your fingers and the rod creates a standing wave that produces a sound. If you have the right amount of friction, the sound gets louder and louder with every stroke.

Where you hold the rod affects both the pitch and loudness of the sound. When holding the rod at a node, the tone is louder, but when holding the rod where an antinode would occur, the sound stops. Moving your hand to different nodes changes the wavelength of the standing wave and therefore the frequency. You can measure the frequencies of different sounds from the rods. You get the fundamental wavelength (i.e., the $N = 1$ or first harmonic) when holding the rod in the middle, in which case the wavelength is equal to twice the length of the rod. If you hold the rod ¼ of the way down, you get the second ($N = 2$) harmonic frequency, which has a wavelength equal to the length of the rod and twice the frequency of the fundamental.

FIGURE 5.10: Music Box

MUSIC BOX

When the Music Box (Figure 5.10) is played while held in the air, its mechanism gets metal wires to vibrate, and the vibrations move molecules of air. Sound waves are created. When the box is on a table, vibrations move from the box to the surface of the table (through the metallic base of the box). The vibrations of the music box cause the surface of the table to start vibrating, and the wider area vibrating makes more air molecules vibrate too; the sound gets louder.

When the music box is on your elbow, the sound travels through your body. Your bones are particularly good conductors of sound, so you hear the music more loudly.

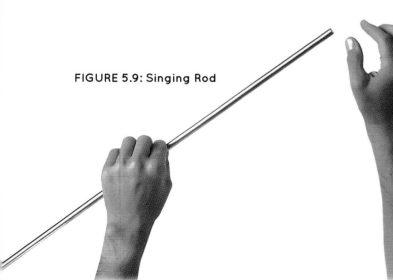

FIGURE 5.9: Singing Rod

FIGURE 5.11: Sound Pipe

SOUND PIPE

In this exploration, you twirl a long, flexible tube called a Sound Pipe (Figure 5.11). Sound is created if air moves fast enough through the pipe. When the pipe is twirled, air molecules move toward the end of the pipe as a result of the centrifugal reaction on air molecules in the rotating tube. When the air flows through the pipe, there are some deviations in the airflow. The deviations, which are usually referred to as *turbulence*, occur because the pipe's surface is not smooth. The pressure waves from the turbulence are partially reflected from the open pipe ends, and the waves reinforce each other for those frequencies that are resonant frequencies. This would not work if the wavelength of the reflected wave were not the same as the wavelength of the incoming wave. Since it *is* the same wavelength, the waves can constructively interfere with each other. None of this would occur if the pipe were smooth—that is, with no corrugations.

Which wavelengths get reinforced depends on the length of the pipe. If the wavelength is twice the length of the pipe, equal to the length of the pipe, 2/3 the length of the pipe, half the length of the pipe, and so on, the reflected sound waves will intensify. In the pipe, there are nodes and antinodes. The lowest frequency that you hear when twirling the pipe slowly is the second, $N = 2$, harmonic. (The fundamental isn't heard because if you twirl the pipe slowly enough for the fundamental frequency, you don't get enough turbulence in the tube to make an audible sound.) As the tube is twirled faster, higher harmonics can be heard.

Extension

Create your own musical instrument.

Web Resources

See how reflected waves result in a standing wave. Vary the wavelength.
www.colorado.edu/physics/2000/microwaves/standing_wave2.html

Explore standing waves on a string. You can vary the frequency, amplitude, and length of the string.
http://ngsir.netfirms.com/englishhtm/StatWave.htm

See both standing waves and traveling waves, transverse waves and longitudinal waves. Select the harmonic number. See where the nodes and antinodes are.
http://ngsir.netfirms.com/englishhtm/TwaveStatA.htm

Specify the length of a tube (open or closed) and see the resonant frequencies. See how it changes with changes in air temperature.
http://hyperphysics.phy-astr.gsu.edu/hbase/waves/opecol.html

Make waves on a string. To make standing waves, click on "Oscillate" and set the damping to zero. Adjust the frequency and the string tension.
http://phet.colorado.edu/sims/wave-on-a-string/wave-on-a-string_en.html

See in slow motion how waves traveling in opposite directions can constructively interfere and create a standing wave. Click the ">" (play) button on the far right to start.
www2.biglobe.ne.jp/~norimari/science/JavaEd/e-wave4.html

Animation of resonance in a tube.
www.sciencejoywagon.com/physicszone/otherpub/wfendt/stlwaves.htm

An explanation of resonance in tubes (both open and closed).
http://dev.physicslab.org/Document.aspx?doctype=3&filename=WavesSound_ResonancePipes.xml

Relevant Standards

Note: The Next Generation Science Standards *can be viewed online at* www.nextgenscience.org/next-generation-science-standards.

PERFORMANCE EXPECTATIONS

HS-PS4-1

Use mathematical representations to support a claim regarding relationships among the frequency, wavelength, and speed of waves traveling in various media.

HS-PS4-5

Communicate technical information about how some technological devices use the principles of wave behavior and wave interactions with matter to transmit and capture information and energy.

SCIENCE AND ENGINEERING PRACTICES

Developing and Using Models

Modeling in 9–12 builds on K–8 and progresses to using, synthesizing, and developing models to predict and show relationships among variables between systems and their components in the natural and designed worlds.

- Develop and use a model based on evidence to illustrate the relationships between systems or between components of a system.

Planning and Carrying Out Investigations

Planning and carrying out investigations to answer questions or test solutions to problems in 9–12 builds on K–8 experiences and progresses to include investigations that provide evidence for and test conceptual, mathematical, physical, and empirical models.

- Plan and conduct an investigation individually and collaboratively to produce data to serve as the basis for evidence, and in the design: decide on types, how much, and accuracy of data needed to produce reliable measurements and consider limitations on the precision of the data (e.g., number of trials, cost, risk, time), and refine the design accordingly.

Using Mathematics and Computational Thinking

Mathematical and computational thinking at the 9–12 level builds on K–8 and progresses to using algebraic thinking and analysis, a range of linear and nonlinear functions including trigonometric functions, exponentials and logarithms, and computational tools for statistical analysis to analyze, represent, and model data. Simple computational simulations are created and used based on mathematical models of basic assumptions.

- Create a computational model or simulation of a phenomenon, designed device, process, or system.

Constructing Explanations and Designing Solutions

Constructing explanations and designing solutions in 9–12 builds on K–8 experiences and progresses to explanations and designs that are supported by multiple and independent student-generated sources of evidence consistent with scientific ideas, principles, and theories.

- Design, evaluate, and/or refine a solution to a complex real-world problem, based on scientific knowledge, student-generated sources of evidence, prioritized criteria, and tradeoff considerations.

CONNECTIONS TO NATURE OF SCIENCE

Science Models, Laws, Mechanisms, and Theories Explain Natural Phenomena

- Theories and laws provide explanations in science.

- Laws are statements or descriptions of the relationships among observable phenomena.

DISCIPLINARY CORE IDEAS

PS4.A: Wave Properties

- The wavelength and frequency of a wave are related to one another by the speed of travel of the wave, which depends on the type of wave and the medium through which it is passing.

- Waves can add or cancel one another as they cross, depending on their relative phase (i.e., relative position of peaks and troughs of the waves), but they emerge unaffected by each other.

PS4.C: Information Technologies and Instrumentation

- Multiple technologies based on the understanding of waves and their interactions with matter are part of everyday experiences in the modern world (e.g., medical imaging, communications, scanners) and in scientific research. They are essential tools for producing, transmitting, and capturing signals and for storing and interpreting the information contained in them.

CROSSCUTTING CONCEPTS

Energy and Matter

- Changes of energy and matter in a system can be described in terms of energy and matter flows into, out of, and within that system.

- Energy cannot be created or destroyed—only moves between one place and another place, between objects and/or fields, or between systems.

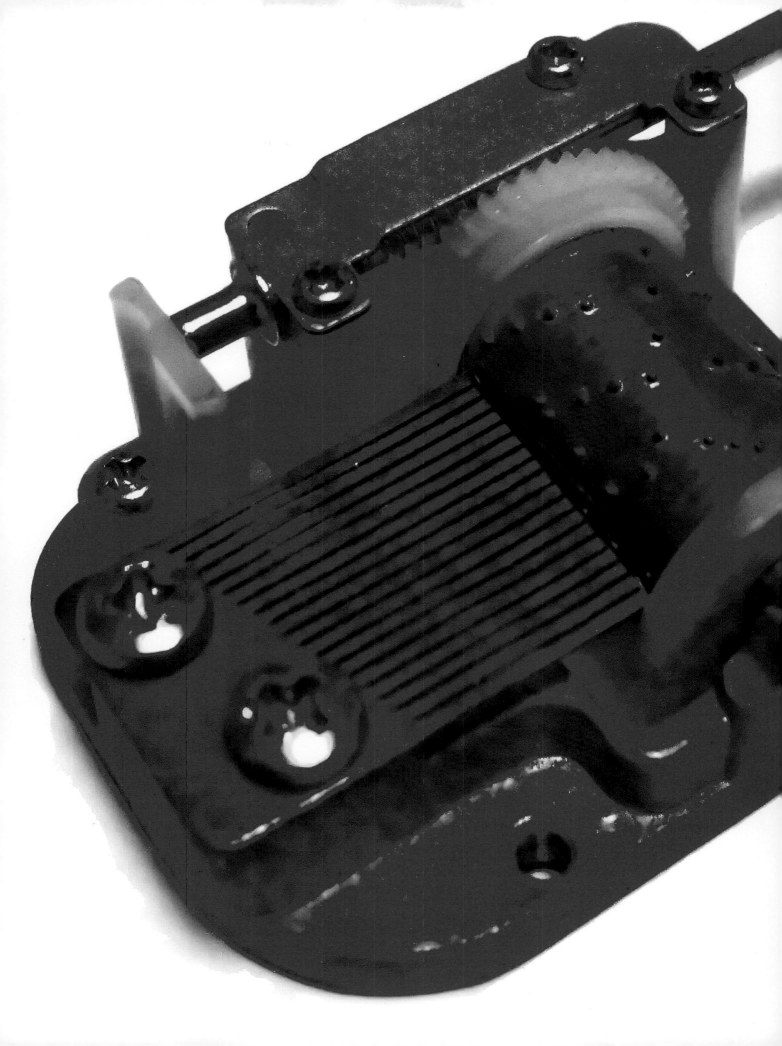

6

BUOYANCY

When an object is in a fluid, such as when a balloon is in the air or a submarine is in the water, a buoyant force acts on it. We first experience a fluid's buoyant force when, as children, we learn to float and swim in water. If our lungs are full of air, floating is easier since inhaling increases our volume, thus lowering our density. Floating in seawater is easier than in freshwater because the salt makes the seawater denser (Figure 6.1). Buoyancy is also responsible for the functioning of floatation devices.

The effect of buoyancy is easily felt in water. For example, it is less difficult to lift heavy objects like stones when they are in water rather than in air. An anchor will feel lighter in water than out of water.

The principle of buoyancy can also be important when objects are in the air. Balloons filled with helium must be attached to a string to prevent them from floating away. Hot air balloons and blimps can lift humans into the air (Figure 6.2) because the helium in the blimp, or the hot air in the balloon, is less dense than the surrounding air.

FIGURE 6.1: Buoyancy keeps boats afloat.

FIGURE 6.2: Buoyancy can produce very strong lifting forces.

FIGURE 6.5: Poly Density Bottles

▼
SAFETY NOTES

- Wear safety glasses or goggles.

- Immediately wipe up any splashed water to prevent a slip or fall hazard.

MIX IT UP

Poly Density Bottles (Figure 6.5) allow us to investigate the interaction between objects and fluid. An object floats in a fluid if its density is less than that of the fluid. If there are many different objects or fluids, the denser objects sink to the bottom of the container and the less dense objects rise to the top.

Procedure

1. Shake the bottle and put it on the table.

2. Watch what happens inside the bottle.

Questions

- What do the different-colored beads do at the beginning of the experiment? Why?

- How do the beads' behaviors change after a while? Why?

- Does the volume of the liquid change during the experiment?

SOLAR BAG

Buoyancy is a phenomenon that occurs not only in water but also in the atmosphere. A hot air balloon rises up in the air because of a buoyant force from the air around it. In this experiment, you see what happens when heat from the sun warms the air inside the Solar Bag (Figure 6.6).

Procedure

1. On a sunny day, bring the Solar Bag outside to an open field.

2. Tie one end of the bag.

3. Fill the bag with air and tie the other end of the bag.

4. Tie a string (5–10 m long) on one end of the bag, and tie the other end of the string to something on the ground.

5. Wait (several minutes or more) until the sun heats the bag and watch what happens.

Questions

• Explain what happens to the Solar Bag.

• Why is cold air denser than hot air?

FIGURE 6.6: Solar Bag

ROCK THE BOAT

A boat floats because the buoyant force equals the weight of the boat. When the boat's weight increases (because you've put something in the boat), it sinks deeper into the water, and the buoyancy increases. If a heavy weight—such as a large rock—is thrown overboard into the water, the buoyancy decreases. What happens to the level of the water then? In this experiment with the Boat & Rock (Figure 6.7), we use a weight to represent the rock.

PART 1

Procedure

1. Measure the mass of the boat and rock together.

2. Place a small container under the tube to catch the water that comes out.

3. With the valve on the overflow tube open, fill the container with water up to the tube (until a little water comes out). Dump out the water that overflowed so that you're starting with an empty overflow container.

4. Put the boat and rock in the water.

5. Measure the mass of the displaced water, and compare it to the mass of the boat and rock.

Questions

- How does the mass of the displaced water compare with the mass of the boat and rock?

- Explain your results.

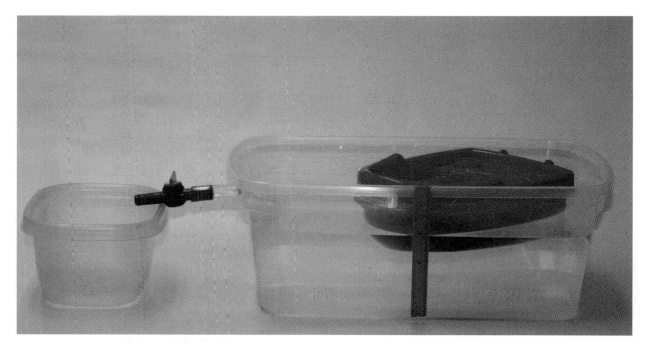

FIGURE 6.7: Boat & Rock

PART 2

Procedure

1. Close the valve and put the boat in the container with water.

2. Place the rock (the weight) in the water and see if it floats.

3. Place the rock in the boat.

4. Mark the level of the water on the container (for example, with a piece of tape).

5. Note the level of the water on the side of the boat. (This is called the *draft*— also spelled draught—of the boat.)

6. See what happens to the boat's draft and the water level in the container when you move the rock from the boat into the water.

7. Repeat the experiment with the wooden weight.

Questions

- How does the water level change when you put the rock (weight) into the boat? Why?

- Why does the draft of the boat change when you put the weight into the boat?

- What happens to the level of the water in the container when moving the weight from the boat to the water? Why?

CARTESIAN DIVER

The Cartesian Diver (Figure 6.8) allows us to investigate the effects of pressure on an object's buoyancy. When squeezing the bottle, the pressure in the bottle increases. The higher pressure compresses the air in the diver, so the volume of air in the diver decreases. With the air in the diver now taking up less space, water flows into the diver, filling the volume no longer occupied by the air. Because of the extra water, the mass of the diver increases. The buoyancy remains the same, but because the diver's weight is now greater than the buoyancy, the diver sinks.

FIGURE 6.8: Cartesian Diver

GALILEO'S THERMOMETER

The cylinder of Galileo's thermometer (Figure 6.9) contains water and a number of weights, each labeled with a temperature. The weights are attached to glass bulbs containing a colored liquid. As the liquid inside the cylinder changes temperature and therefore density, the glass bulbs rise or fall in the liquid. If the density of the glass bulb and weight is greater than density of the liquid at that temperature, the glass bulb sinks. The glass bulbs differ in density by a small amount, so, depending on the temperature, different numbers of bulbs will float. This creates a temperature scale since the least dense bulb is at the top and the densest is at the bottom. Therefore the lowest floating bulb shows the temperature inside the tube.

FIGURE 6.9: Galileo's thermometer

FIGURE 6.10: Poly Density Bottles

MIX IT UP

In the Poly Density Bottle (Figure 6.10) there are white beads and blue beads, alcohol, water, and salt. Alcohol does not dissolve in the salt water, so after shaking the bottle the liquids are disturbed and temporarily mixed but not dissolved. The density of the alcohol–water–salt mixture is less than the density of the blue beads but greater than the density of the white beads. First the white beads rise to the top and the blue beads sink. After a while, when the alcohol starts to separate from the salt water, the white beads sink and the blue beads rise. The salt water is denser than the beads and alcohol. Finally, the less-dense alcohol moves to the top, and the salt water lies at the bottom.

SOLAR BAG

The Solar Bag (Figure 6.11) demonstrates that an object immersed in fluid is buoyed up with a force equal to the weight of the fluid displaced by the object. With the Solar Bag, the sun heats the molecules of air inside the bag, making the pressure rise. The increased pressure pushes outward on the bag, increasing its volume. (The amount or mass of air inside the bag does not increase when the bag's volume grows, so the density decreases.) As the volume increases, the bag displaces more cold air. When the weight of the displaced air is greater than the bag's weight (including the weight of the

FIGURE 6.11: Solar Bag

FIGURE 6.12: Boat & Rock

air inside it), the bag will start to rise into the air. The density of the air outside the Solar Bag is then greater than the density of the air inside the bag.

ROCK THE BOAT

The Boat & Rock (Figure 6.12) activity explores the buoyancy that results from the pressure in a fluid. The weight of fluid makes the pressure higher as you go deeper. For an object in the fluid, the bottom of the object is lower and will therefore have a higher pressure on it than the top of the object. Therefore there will be a greater force upward on the bottom of the object than downward on the top of the object. The *net* force from the water (which is the difference in pressure between the top and bottom) is an upward force—in the opposite direction of gravity. That force is called buoyancy. If an object is floating, the buoyancy equals the weight of the object.

Buoyancy can be calculated with Archimedes' principle: An object immersed in a fluid is buoyed up with a force equal to the weight of the fluid displaced by the object. This means that the displaced

water in the "Rock the Boat" exploration (which overflows through the tube) should have a mass equal to that of the boat plus the rock (or in this case, the weight).

The rock doesn't float because its density is greater than the water's density. The combined boat and rock displace water from the container such that the weight of the displaced water is equal to the weight of the boat and rock. When the rock is moved into the water, two things happen: (1) The rock displaces an amount of water with the same volume as the rock. That would make the water level in the container rise a little. (2) At the same time, the boat rises up in the water (i.e., has less draft) because the weight of the rock is no longer causing additional water to be displaced. This makes the water level go down. So now the question is, which is greater—the amount the water goes up because of (1) or the amount that the water goes down because of (2)? The amount of water no longer displaced (as described in [2]) had a weight equal to the weight of the rock. Since the rock is denser than water, this amount of water—having weight equal to the weight of the rock—must have a greater volume than the rock. So the amount by which the water level goes down due to (2) is greater than the amount the water level rises due to (1), and the net effect is that the water level in the container goes down.

Web Resources

Learn about buoyancy by floating blocks in liquid.
http://phet.colorado.edu/en/simulation/buoyancy

Explore how size, weight, and density affect an object's buoyancy.
http://phet.colorado.edu/en/simulation/density

See how much water can be added to a vessel before it sinks. Monitor the mass, density, and volume as you go.
www.mhhe.com/physsci/physical/giambattista/buoyancy/buoyancy.html

Play with the effects of buoyancy in liquid.
www.walter-fendt.de/ph14e/buoyforce.htm

Experiment with the Cartesian diver to learn about Archimedes' principle as well as the ideal gas law.
http://lectureonline.cl.msu.edu/~mmp/applist/f/f.htm

Experiment with Galileo's thermometer, where you estimate the temperature by looking at the colored beads. You can modify the temperature of the liquid as well as the minimum and maximum densities of the bulbs.
http://scratch.mit.edu/projects/27574/

Relevant Standards

Note: The Next Generation Science Standards *can be viewed online at* www.nextgenscience.org/next-generation-science-standards.

SCIENCE AND ENGINEERING PRACTICES

- Plan and conduct an investigation individually and collaboratively to produce data to serve as the basis for evidence, and in the design: decide on types, how much, and accuracy of data needed to produce reliable measurements and consider limitations on the precision of the data (e.g., number of trials, cost, risk, time), and refine the design accordingly.

- Analyze data using tools, technologies, and/or models (e.g., computational, mathematical) in order to make valid and reliable scientific claims or determine an optimal design solution.

- Constructing explanations and designing solutions in 9–12 builds on K–8 experiences and progresses to explanations and designs that are supported by multiple and independent student-generated sources of evidence consistent with scientific ideas, principles, and theories.

CONNECTIONS TO NATURE OF SCIENCE

Science Models, Laws, Mechanisms, and Theories Explain Natural Phenomena

- Theories and laws provide explanations in science.

- Laws are statements or descriptions of the relationships among observable phenomena.

CROSSCUTTING CONCEPTS

Patterns

- Different patterns may be observed at each of the scales at which a system is studied and can provide evidence for causality in explanations of phenomena.

Cause and Effect

- Empirical evidence is required to differentiate between cause and correlation and make claims about specific causes and effects.

- Systems can be designed to cause a desired effect.

TWO-DIMENSIONAL MOTION

Two-dimensional (2D) motion means motion that takes place in two different directions (or coordinates) at the same time. The simplest motion would be an object moving linearly in one dimension. An example of linear movement would be a car moving along a straight road or a ball thrown straight up from the ground.

If an object is moving in one direction with a constant velocity while accelerating in another direction, calculating the motion is more complicated. An example of 2D movement would be throwing a football or hitting a home run in baseball. In the following examples, we explore motion in the Earth's gravitational field.

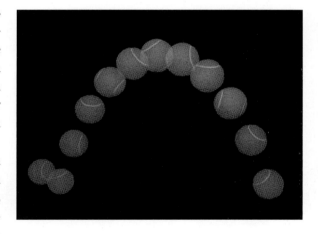

7E Exploration

RACE TO THE BOTTOM

In this experiment, you will use the Vertical Acceleration Demonstrator (Figure 7.1) to explore what effect horizontal motion has on the vertical acceleration of a falling object.

Procedure

1. Use a clamp to attach the apparatus to a ring stand so that the height of the apparatus is ~1.5 meters.

2. Mount the two balls on the apparatus, making sure that the apparatus is level.

3. Release the spring and watch as one ball is projected horizontally and the other is dropped straight down. Listen to them hit the floor to determine whether there is a difference in when the balls reach the ground.

Questions

- Which ball hit the floor first? Why?

- What forces are involved?

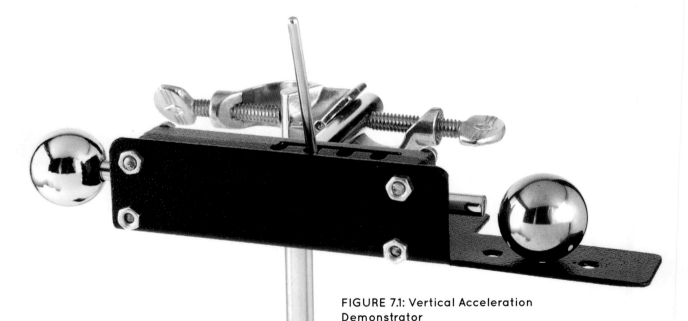

FIGURE 7.1: Vertical Acceleration Demonstrator

EJECTOR SEAT

With the Ballistics Car (Figure 7.2), you will explore relative motion and use different coordinates to describe movement.

Procedure

1. Discuss with your group where the ball will land if it is launched straight up from the barrel while the car is moving.

2. Place the ball on the piston inside the barrel.

3. Push the piston down and lock it in place with the pin.

4. Push the car away from you while holding the string from the pin in your hand.

5. Launch the ball by giving the string a sharp pull.

Questions

• Where does the ball land?

• What are the forces affecting the ball after its launch?

• Does the velocity of the car affect what happens?

• Explain your results.

▼
SAFETY NOTES

• Wear safety glasses or goggles.

• Make sure the trajectory path of the launched ball is clear of observers, fragile equipment, and so on, to prevent damage or injury.

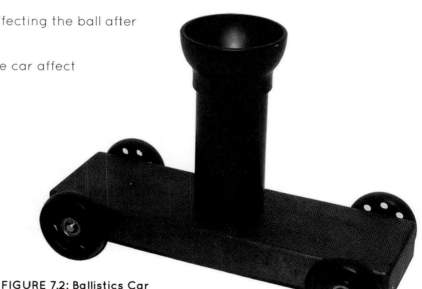

FIGURE 7.2: Ballistics Car

NEED FOR SPEED

You might recognize this car track (Figure 7.3) if you
did the Energy Lab in Chapter 3. We're going to use it
to demonstrate 2D motion here. In many movies, a car
or motorcycle is used to jump over a river or from one
building to another. To succeed with the jump, accurate
calculations have to be made. In the first procedure in
this exploration, the car will be launched horizontally, and
in the second procedure it will be launched at an incline.

Procedure 1: Jumping Over the Ravine

1. Build a ravine at your table (i.e., create a gap for your
    car to jump over, for example, from the table to your
    chair).

2. Launch the car so it goes down the ramp and ends
    up moving horizontally. Measure the velocity with the
    photogate.

3. Place the car track on the edge of the ravine.

4. Calculate how far the car could travel with the speed
    you measured.

5. Move the chair to the distance you calculated,
    and test to see how close your calculation was.

6. If more speed is needed, experiment with
    different ways of launching the car (for instance,
    with the aid of a rubber band).

Procedure 2: Curvilinear Motion

1. Build a ramp with the car track on one table and
    a ramp for the landing on another table.
    (The ramp that the car is launched
    from should be inclined at some
    angle.)

2. Calculate the velocity of the car on the ramp based on the velocity measured before.

3. Move the ramp to the edge of the ravine (e.g., edge of the table), and calculate how far the car will travel from the ramp with that velocity.

4. Move the other table to the distance you calculated, and test if the car can make the jump.

Questions

- What's the farthest you can move the chair (or landing platform) on the basis of your calculations?

- What is the distance between the tables in the second experiment?

- Explain why there is a difference between theory (the calculated value) and experiment (what actually happened)?

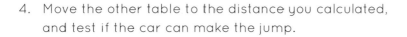

FIGURE 7.3: Car track

FIGURE 7.4: Air-Powered Projectile

▼
SAFETY NOTES

- Wear safety glasses or goggles.

- Make sure the trajectory path of the launched rocket is clear of observers, fragile equipment, and so on, to prevent damage or injury.

LAUNCH AND LAND

In this experiment, an Air-Powered Projectile (Figure 7.4) is launched twice. The first launch is straight up in the air, and the second launch is at an angle up and out toward the field. Work in groups and compete to see which group can most accurately calculate the rocket's landing spot. The calculations for the second launch will make use of data from the first launch. The first (vertical) launch will be used to determine the launch velocity of the rocket.

Procedure

1. Assume that the rocket will have the same initial speed for each launch.

2. Obtain a timer and protractor.

3. Find a good place for the launches, such as a field where there is at least 50 meters of open space. The more space, the better.

4. Attach the rocket to the launch pad.

5. Attach the pump to the rocket; add the plastic cap, and put the rest of the rocket in place.

6. Pump until the rocket is launched. Launch the rocket vertically the first time.

7. From your measurements, calculate the rocket's launch velocity.

8. Next, calculate where the rocket should land if you launch it at a certain angle. Mark the landing spot.

9. Launch the rocket at that angle. See how accurate your calculation was. Discuss possible sources of error.

10. To make the problem more challenging, the rocket can be launched from a higher level than the field.

Questions

- What did you calculate for the landing spot? (Show your calculation.)

- How far off was your calculated landing spot from the actual landing spot?

- Explain possible causes for any discrepancy.

RACE TO THE BOTTOM

In this exploration, you use the Vertical Acceleration Demonstrator (Figure 7.5) to launch two balls—one vertically and the other horizontally. After releasing the spring, the balls are launched from the apparatus and the Earth's gravitational force pulls them toward the ground. One ball is projected horizontally, and the other one starts from rest (zero velocity). The gravitational force is the same for both balls, so the acceleration due to gravity is also the same for both. The balls attain the same vertical velocity toward the ground because they experience the same downward acceleration. Therefore, the ball projected horizontally reaches the floor at the same time as the other ball.

FIGURE 7.6: Ballistics Car

EJECTOR SEAT

The Ballistics Car (Figure 7.6) launches a ball straight up as it moves forward. When the ball is launched, the ball and the car are initially moving in the same direction at the same velocity. The ball is launched vertically straight up in the air. While the ball is in the air, the car keeps on moving in the same direction and with the same velocity as before. (Its velocity does slow down a little bit because of friction and air resistance.) The ball's vertical movement doesn't affect its horizontal movement, so the ball continues to have the same horizontal speed as the car, even while the ball is also moving vertically in the air. The forces affecting the ball are the gravitational force, which pulls the ball back down, and air resistance, which slows the ball's horizontal movement approximately the same amount as the car is slowed by its resistance forces. Because the car and the ball have the same horizontal speed, the ball lands back in the car's barrel. This works whether the car is moving fast or slowly; its velocity doesn't matter as long as it is constant.

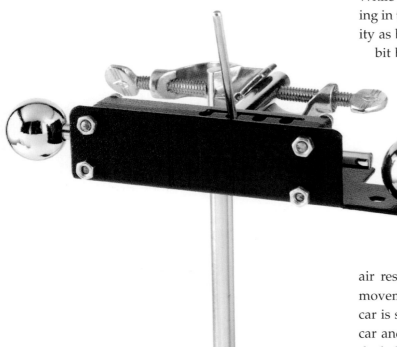

FIGURE 7.5: Vertical Acceleration Demonstrator

NEED FOR SPEED

This exploration uses the same car track (Figure 7.7) as the Energy Lab in Chapter 3. In part 1, you mimic the action of a car jumping a ravine by creating a jump between, for example, a table and chair. In part 2, this study was extended by adding ramps for the car on each end of its journey. The distance that the car travels through the air depends on the car's initial velocity and the difference in height between the table and the chair. Let's call the velocity (v) of the car on the edge of the ravine v_0 and the height difference h; g stands for the acceleration due to gravity, t for time, and s for distance in meters.

FIGURE 7.7: Car track

$$h = \tfrac{1}{2}gt^2$$

so

$$t = \sqrt{\frac{2h}{g}}$$

When the car is falling, it moves a horizontal distance:

$$s = v_0 t = v_0 \sqrt{\frac{2h}{g}}$$

The distance from the ramp can be calculated as $v_x = v_0 \cos \alpha$ and $v_y = v_0 \sin\alpha$, where α is the angle from the horizontal launch. (We define the horizontal direction to be the x direction and the vertical direction to be the y direction.)

Velocities:
vertical $v_y = v_0 \sin\alpha - gt$
horizontal $v_x = v_0 \cos\alpha$

The vertical position, s_y, is then

$$s_y = h + v_0 \sin \alpha \, t - \tfrac{1}{2}gt^2$$

To calculate the flight time, set s_y equal to zero:

$$0 = \tfrac{1}{2}gt^2 - v_0 \sin \alpha \, t - h$$

Then, use the quadratic formula to solve for t:

$$t = \frac{v_0 \sin \alpha \pm \sqrt{(v_0 \sin \alpha)^2 + 2gh}}{2 \cdot \tfrac{1}{2}g}$$

When the time for the jump is known, the distance can be calculated from: $s_x = v_0 \cos\alpha t$

Note that we are ignoring air resistance and the rotation of the car during the jump. Because of these factors, the car's jump is not as long as you calculate.

LAUNCH AND LAND

In this exploration, you launch the Air-Powered Projectile (Figure 7.8) at varying angles and determine the rocket's velocity. Here are two ways to determine the rocket's launch velocity: The first way is to use geometry and trigonometry. The first launch will be straight up. From a known distance from the launch pad, measure the angle (α) up to the rocket's maximum height. Next, use trigonometry to calculate the height, h. Then use that height and conservation of energy to calculate the launch velocity, v_0,

FIGURE 7.8: Air-Powered Projectile

$$mgh = \tfrac{1}{2}mv^2$$

where m is mass and g is the acceleration due to gravity.

The second way to determine the rocket's velocity is to measure flight time. It can be measured starting when the rocket is at its highest point and ending when rocket is back on the ground.

When the rocket is at its highest point, its velocity is zero. $v_y = v_0 - gt$, so when v_y is zero, $v_0 = gt$ (where t is time).

Now suppose you launch the rocket at an initial angle θ. The x and y components of the initial velocity (which are independent of each other) are then given by

$$v_{0x} = v_0 \cos\theta$$

$$v_{0y} = v_0 \sin\theta$$

The horizontal component of velocity, v_x, is constant, so the x position at any time is given by

$$x = v_x t = (v_0 \cos\theta)t.$$

The vertical component of velocity, v_y, changes according to

$$v_y = v_{0y} - gt \text{ or } v_y = v_0 \sin\theta - gt.$$

The rocket's displacement in the vertical direction is given by

$$y = -\tfrac{1}{2}gt^2 + v_{0y}t.$$

Since y is the difference between the initial height and final height, the final height is equal to $h + y$. You can solve this for the flight time, t, by setting $y = 0$ if the rocket is launched from the same level as the field, or, if not, setting y equal to the height that the field is above or below the height of the launch pad. (If the field is at a lower height, then your value of y will be negative.)

Now that you know the time, the distance that the rocket travels horizontally can be calculated from $x = v_x t = (v_0 \cos\theta)t$.

One possible cause of a discrepancy between the theoretical distance and the experimental distance is that the calculation neglected air resistance (and any wind that might be present).

Web Resources

Fire various objects out of a cannon. Set the angle, initial speed, and mass. Add air resistance. Make a game out of this simulation by trying to hit a target.
http://phet.colorado.edu/sims/projectile-motion/projectile-motion_en.html

Tasks related to projectile motion and using relevant quadratic equations.
http://phet.colorado.edu/files/activities/3491/12-14-11%20quadratic%20function%20related%20tasks.docx

Motion in 2D: Learn about position, velocity, and acceleration vectors.
http://phet.colorado.edu/en/simulation/motion-2d

Activity sheets to go with the previous 2D simulation.
http://phet.colorado.edu/files/activities/3595/C:_Users_patrick.foley_Documents_Motion_in_2D.doc
http://phet.colorado.edu/files/activities/2930/Phet%202D%20Motion%20Activity.doc
http://phet.colorado.edu/files/activities/3141/Vectors%20PhET%20Lab.doc

Projectile simulation: Set the angle, initial velocity, and mass of the projectile, and then see its path.
http://galileo.phys.virginia.edu/classes/109N/more_stuff/Applets/ProjectileMotion/jarapplet.html

A more sophisticated projectile simulator. Adjust the initial height and the value of *g*, see a graph of the position, velocity, acceleration, force, and energy, and pause at any point and examine the values of the parameters.
www.walter-fendt.de/ph14e/projectile.htm

Relevant Standards

Note: The Next Generation Science Standards *can be viewed online at* www.nextgenscience.org/next-generation-science-standards.

PERFORMANCE EXPECTATIONS

HS-PS3-1

Create a computational model to calculate the change in the energy of one component in a system when the change in energy of the other component(s) and energy flows in and out of the system are known. [Clarification Statement: Emphasis is on explaining the meaning of mathematical expressions used in the model.] [Assessment Boundary: Assessment is limited to basic algebraic expressions or computations; to systems of two or three components; and to thermal energy, kinetic energy, and/or the energies in gravitational, magnetic, or electric fields.]

HS-PS3-3

Design, build, and refine a device that works within given constraints to convert one form of energy into another form of energy. [Integrates science and engineering]

SCIENCE AND ENGINEERING PRACTICES

Developing and Using Models

Modeling in 9–12 builds on K–8 and progresses to using, synthesizing, and developing models to predict and show relationships among variables between systems and their components in the natural and designed worlds.

- Develop and use a model based on evidence to illustrate the relationships between systems or between components of a system.

Planning and Carrying Out Investigations

Planning and carrying out investigations to answer questions or test solutions to problems in 9–12 builds on K–8 experiences and progresses to include investigations that provide evidence for and test conceptual, mathematical, physical, and empirical models.

- Plan and conduct an investigation individually and collaboratively to produce data to serve as the basis for evidence, and in the design: decide on types, how much, and accuracy of data needed to produce reliable measurements and consider limitations on the precision of the data (e.g., number of trials, cost, risk, time), and refine the design accordingly.

Analyzing and Interpreting Data

Analyzing data in 9–12 builds on K–8 and progresses to introducing more detailed statistical analysis, the comparison of data sets for consistency, and the use of models to generate and analyze data.

- Analyze data using tools, technologies, and/or models (e.g., computational, mathematical) in order to make valid and reliable scientific claims or determine an optimal design solution.

Using Mathematics and Computational Thinking

Mathematical and computational thinking at the 9–12 level builds on K–8 and progresses to using algebraic thinking and analysis, a range of linear and nonlinear functions including trigonometric functions, exponentials and logarithms, and computational tools for statistical analysis to analyze, represent, and model data. Simple computational simulations are created and used based on mathematical models of basic assumptions.

- Create a computational model or simulation of a phenomenon, designed device, process, or system.

Constructing Explanations and Designing Solutions

Constructing explanations and designing solutions in 9–12 builds on K–8 experiences and progresses to explanations and designs that are supported by multiple and independent student-generated sources of evidence consistent with scientific ideas, principles, and theories.

- Design, evaluate, and/or refine a solution to a complex real-world problem, based on scientific knowledge, student-generated sources of evidence, prioritized criteria, and tradeoff considerations.

DISCIPLINARY CORE IDEAS

PS2.B: Types of Interactions

- Newton's law of universal gravitation and Coulomb's law provide the mathematical models to describe and predict the effects of gravitational and electrostatic forces between distant objects.

- Forces at a distance are explained by fields (gravitational, electric, and magnetic) permeating space that can transfer energy through space. Magnets or electric currents cause magnetic fields; electric charges or changing magnetic fields cause electric fields.

PS3.C: Relationship Between Energy and Forces

- When two objects interacting through a field change relative position, the energy stored in the field is changed.

CONNECTIONS TO NATURE OF SCIENCE

Science Models, Laws, Mechanisms, and Theories Explain Natural Phenomena

- Theories and laws provide explanations in science.

- Laws are statements or descriptions of the relationships among observable phenomena.

CROSSCUTTING CONCEPTS

Cause and Effect

Empirical evidence is required to differentiate between cause and correlation and make claims about specific causes and effects.

Systems and System Models

When investigating or describing a system, the boundaries and initial conditions of the system need to be defined.

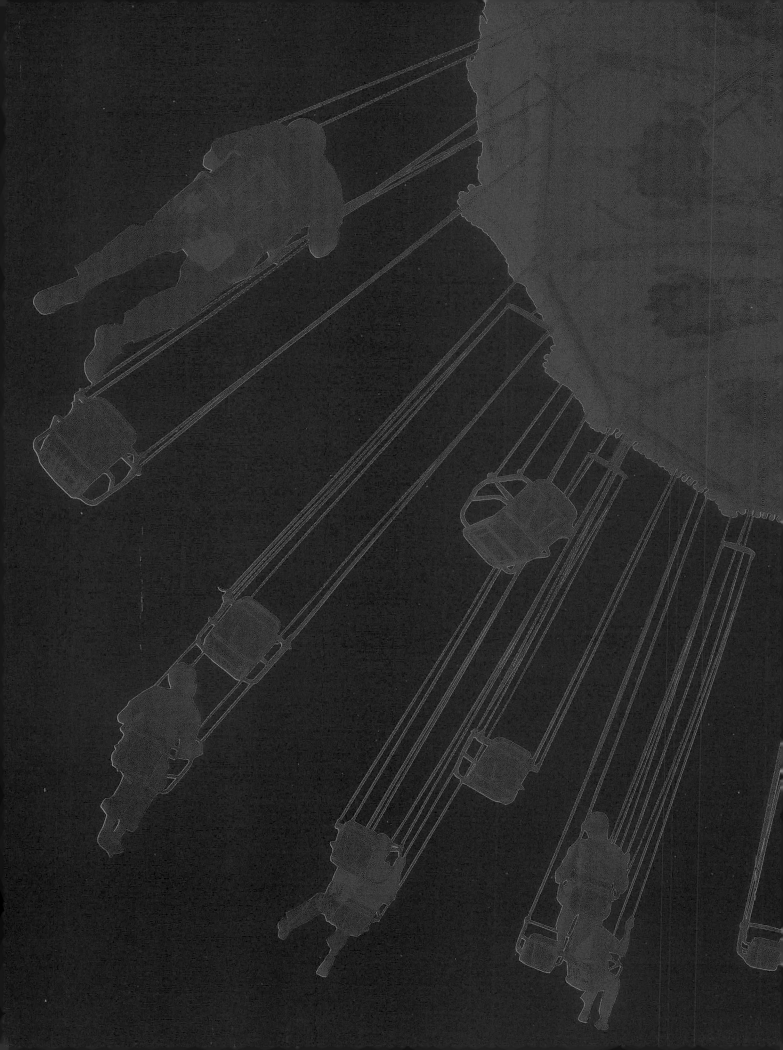

8

ANGULAR MOMENTUM

The concept of angular momentum can be related to linear momentum. The difference is that while linear momentum occurs with motion in a straight line, angular momentum applies to circular motion. Linear momentum is calculated as the product of mass and velocity, whereas angular momentum is the product of moment of inertia and angular velocity. Angular velocity, ω, is the angle swept out by the radius per unit time. The units of angular velocity are radians per second, abbreviated rad/s (or sometimes just s^{-1}).

The moment of inertia (symbolized by I) is a quantity that describes the inertia of the rotation. It tells you how easy it is to change the rate of rotation.

Exploration

▼
SAFETY NOTES

- Wear safety glasses or goggles.

- Use caution when rotating. Do not drop the weights or rod—there is the potential for foot injury.

- Make sure area is clear of furniture, equipment, and so on, in case a student loses his or her balance and falls.

SPEED SPINNING

Perhaps you are familiar with conservation of momentum in the case of linear motion. In the following experiment, you will use a Rotating Platform (Figure 8.1) to investigate the conservation of angular momentum.

Procedure

1. Hold a 1-kg weight in each hand and step on the rotating platform.

2. Spread your arms and then have someone gently start you rotating.

3. Once you are rotating, pull your hands close to your body.

4. Try it again, this time while holding the heavy metal rod. Hold the rod horizontally at the beginning, and while you are rotating turn the rod upright.

FIGURE 8.1: Rotating Platform

Questions

- Why does your angular velocity change when you pull your arms close to your body?

- Why did turning the rod change your angular velocity?

- Learn what "cross product" means and how angular velocity is defined.

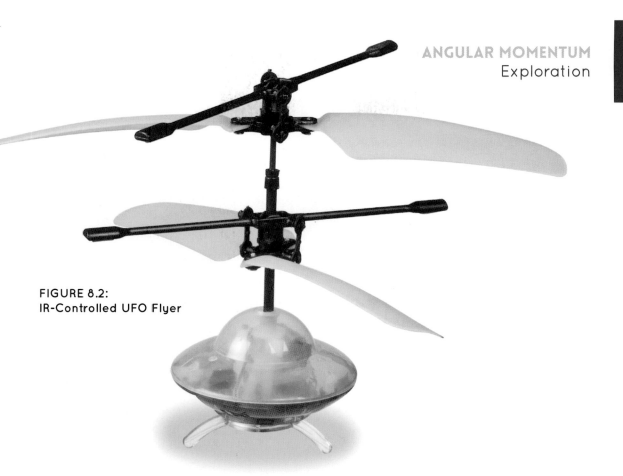

FIGURE 8.2:
IR-Controlled UFO Flyer

PROPELLER PUZZLE

In an isolated system, the angular momentum is constant
if there is not any external torque. The law of conservation
of angular momentum tells us that rotational motion
does not arise from nothing. Use the IR-Controlled
UFO Flyer (Figure 8.2) to see this theory in practice.

Procedure

1. Hold the flyer in your hand and turn on the engine.

2. See which way the rotors are turning.

3. Release the flyer and see which way the rotors turn.

Questions

*   When holding the flyer in your hand, which way are the
    rotor(s) turning.

*   Does the direction the rotors are turning change when
    you release the flyer from your hand? In what way?

*   Is the flyer itself rotating?

▼
SAFETY NOTES

*   Wear safety glasses or
    goggles.

*   Make sure the trajectory
    path of the flyer is clear
    of observers, fragile
    equipment, and so on,
    to prevent damage or
    injury.

FIGURE 8.3: Power Ball With Speed Meter

SAFETY NOTE

Make sure area is clear of observers, fragile equipment, and so on, to prevent damage or injury.

POWER BALL

The Power Ball with Speed Meter (Figure 8.3) demonstrates some effects of torque. The change in the angular momentum (L) per unit time (t) equals the torque, τ:

$$\tau = \frac{\Delta L}{\Delta T}$$

Changing the direction of the axis of rotation requires a torque.

Procedure

1. Start the rotor inside the ball, rotating it either manually or with a motorized starter.

2. Start to rotate your wrist in a slow circular motion to make the rotor rotate faster.

3. When the rotor is rotating fast, turn the gyro ball upside down.

4. Try rolling the gyro ball on the floor.

Questions

• Why is it hard to turn the gyro ball when the rotor is rotating?

• Can you roll the gyro ball on the floor? Why or why not?

WOBBLY TOP

While the Extreme Gyroscope (Figure 8.4) is spinning on the table, it is also wobbling. This wobbling movement is called *precession*. Precession plays an important role in astronomy. For example, Earth both rotates about its axis and precesses. The Earth's precession will cause the seasons to occur in different parts of the Earth's orbit in the distant future. (However, the calendar dates of the seasons will remain approximately the same.) In addition to precession, the gyroscope shows another small motion called *nutation*.

FIGURE 8.4: Extreme Gyroscope

Procedure 1: Precession

1. Start the gyroscope spinning by pulling the string or the T-handle.

2. Place the gyroscope on the string horizontally (see Figure 8.5).

3. Holding the gyroscope in the air with the string, gently try to tip it over.

String

4. Watch what happens when you allow the gyroscope to move freely.

FIGURE 8.5: Gyroscope

5. You can repeat the experiment with the rotating gyroscope

   - on its plastic stand on the table,

   - on a pen, or

   - moving along a string.

Procedure 2: Precession and Nutation

1. Start two gyroscopes rotating.

2. Place the first gyroscope down horizontally and the second one vertically on the first gyroscope.

3. Release the gyroscopes and observe how they "dance."

Questions

- What forces act on the gyroscope while it is hanging on the string?

- Why does the Earth's gravitational force not cause the gyroscope to fall?

- What happens to the gyroscope's axis of rotation direction while it's on the string? Why?

- What creates nutation?

- How are precession and nutation observed in the procedure with two gyroscopes?

▼
SAFETY NOTES

- Wear safety glasses or goggles.

- Make sure the area is clear of observers, fragile equipment, and so on, to prevent damage or injury

Relevant Standards

Note: The Next Generation Science Standards *can be viewed online at* www.nextgenscience.org/next-generation-science-standards.

PERFORMANCE EXPECTATIONS

HS-PS2-2

Use mathematical representations to support the claim that the total momentum of a system of objects is conserved when there is no net force on the system. [Clarification Statement: Emphasis is on the quantitative conservation of momentum in interactions and the qualitative meaning of this principle.]

SCIENCE AND ENGINEERING PRACTICES

Planning and Carrying Out Investigations

Planning and carrying out investigations to answer questions or test solutions to problems in 9–12 builds on K–8 experiences and progresses to include investigations that provide evidence for and test conceptual, mathematical, physical, and empirical models.

- Plan and conduct an investigation individually and collaboratively to produce data to serve as the basis for evidence, and in the design: decide on types, how much, and accuracy of data needed to produce reliable measurements and consider limitations on the precision of the data (e.g., number of trials, cost, risk, time), and refine the design accordingly.

Using Mathematics and Computational Thinking

Mathematical and computational thinking at the 9–12 level builds on K–8 and progresses to using algebraic thinking and analysis, a range of linear and nonlinear functions including trigonometric functions, exponentials and logarithms, and computational tools for statistical analysis to analyze, represent, and model data. Simple computational simulations are created and used based on mathematical models of basic assumptions.

- Create a computational model or simulation of a phenomenon, designed device, process, or system.

Constructing Explanations and Designing Solutions

Constructing explanations and designing solutions in 9–12 builds on K–8 experiences and progresses to explanations and designs that are supported by multiple and independent student-generated sources of evidence consistent with scientific ideas, principles, and theories.

- Design, evaluate, and/or refine a solution to a complex real-world problem, based on scientific knowledge, student-generated sources of evidence, prioritized criteria, and tradeoff considerations.

CONNECTIONS TO NATURE OF SCIENCE

Science Models, Laws, Mechanisms, and Theories Explain Natural Phenomena

- Theories and laws provide explanations in science.

- Laws are statements or descriptions of the relationships among observable phenomena.

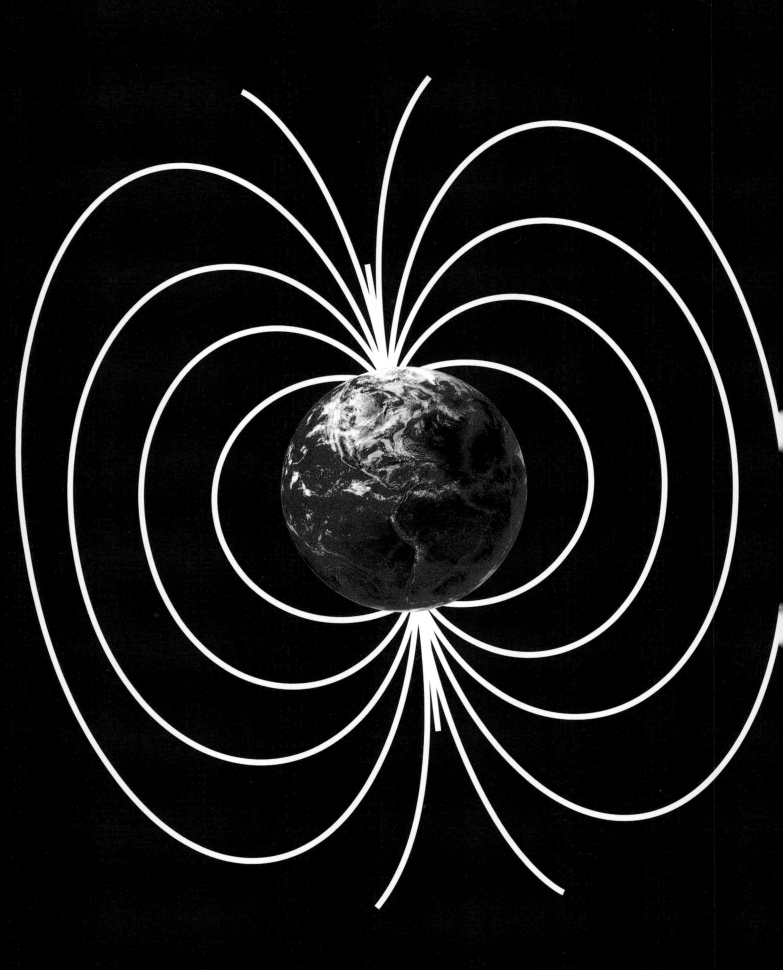

9

MAGNETISM

In ancient Greece, a new mineral was discovered that attracted iron and steel. This mineral was named *magnetite* after the region where it was found, Magnesia. The word magnet comes from a Greek word meaning "Magnesian stone." However, the phenomenon we call magnetism was observed in China long before the Greeks found it. A stone that contains magnetite is called *lodestone*, and the Chinese used lodestone to make a mariner's compass some 4,500 years ago.

Nowadays, magnetism is a well-known physics phenomenon that is used, for example, in computers for magnetic recording, in Maglev trains, or for holding notes on a refrigerator door. There is a magnetic field around the Earth, and one of the oldest applications of magnetism is the compass, which has been used for navigation for a couple thousand years. The compass needle turns parallel to the Earth's magnetic field lines.

Magnetism is a phenomenon in which the magnetic materials interact with one another by forces that attract or repel. A *magnetic field* is a region in which there is a magnetic interaction. Items that produce these magnetic effects are called *magnets*. A magnet always has two magnetic poles, a north pole and a south pole. The north poles of two magnets repel each other, while a north pole and a south pole attract each other.

Exploration

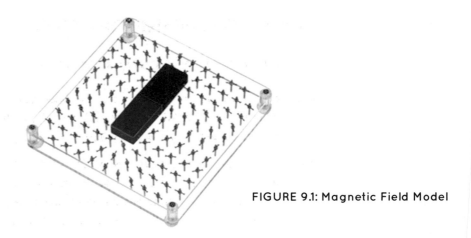

FIGURE 9.1: Magnetic Field Model

▼
SAFETY NOTE

Wear safety glasses or goggles.

POINT THE WAY

Every magnet is surrounded by a magnetic field. With the Magnetic Field Model (Figure 9.1), you can explore the shapes and strengths of magnetic fields from a bar magnet and, if you have it, the Levitron.

Procedure

Observe what happens after each of the following steps:

1. Place a bar magnet on the top plate.

2. Place two bar magnets on the plate so that the magnets repel each other.

3. Place a bar magnet on the plate oriented upward.

4. If you have the Levitron, examine its magnetic field.

Questions

• What is magnetism?

• How does the magnetic field model device work?

• Sketch the magnetic fields produced by each of the different types of magnets that you tried.

• Find out about paramagnetism, diamagnetism, and ferromagnetism. Which of these would apply to the material between the plates?

FIGURE 9.2: Small Clear Compass

MAGNETS IN MOTION

Within a compass like the Small Clear Compass (Figure 9.2) there is a sensitive movable magnetic needle that is balanced at the center. The needle can rotate in the horizontal plane and will rotate to align with the magnetic north-south direction.

A magnetic field is always generated around a moving charged particle. An electric current is simply the movement of charged particles, such as electrons moving through a conductor. Therefore a magnetic field is generated around the conductor when there is an electric current (Figure 9.3). In addition to the compasses, you will need a long wire that can be connected to a battery.

▼
SAFETY NOTE
Wear safety glasses or goggles.

FIGURE 9.3: Compasses around a bar magnet

Procedure

1. Place the small compasses on the table at regular intervals.

2. Place the bar magnet on the plate, and examine the magnetic field with compasses.

3. If some compasses are too far away from the magnet to see an effect, move them closer.

4. Turn the permanent magnet 180° so that the magnetic poles are interchanged. Observe any changes in the compasses.

5. Remove the bar magnet and make a simple electric circuit with a wire and a battery. Place the wire on one of the compasses. Observe the compass.

6. Reverse the direction of the current and observe any change to the compass.

7. Place other compasses near the wire.

Questions

• What shape is the magnetic field formed around the permanent magnet? Pay attention to the magnitude and direction of the magnetic field lines.

• How does the electric current influence the compasses?

• Explain in which direction the magnetic field from a wire is oriented.

FIGURE 9.4: 3D Magnetic Compass

ALL ENCOMPASSING

The 3D Magnetic Compass (Figure 9.4)
has a gimbal-mounted bar magnet in
its head that can turn freely. This device
can be used to observe the direction of a
magnetic field in three dimensions. The
strength of the magnetic field can also be
detected using this probe. You can explore
the magnetic field from current in a wire
or from a permanent magnet such as the
one in the Levitron. In addition to the 3D
Magnetic Compass, you will need a long
wire that can be connected to a battery.

Procedure

1. Place a wire in between two posts,
    stands, or any supports, so that it's
    hanging in the air. This will allow you
    to explore the magnetic field from the
    wire on all sides of it.

2. Make a simple electric circuit with a
    wire and battery, and investigate the
    magnetic field around the conductor
    using the 3D Magnetic Compass.

3. Change the direction of the current
    and investigate how the magnetic field
    changes.

4. If you have the Levitron, take a close
    look at what kind of magnetic field it
    creates around itself. From that, try to
    figure out what is inside the Levitron.

Questions

*   How does the amount of current affect
    the strength of the magnetic field?

*   How does the distance from the wire
    affect the magnetic field strength?

*   If you were able to do step 4, draw the
    Levitron's magnetic field.

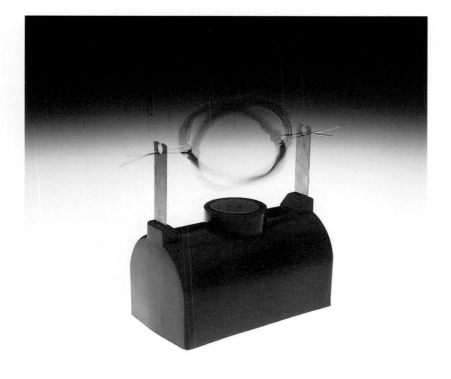

FIGURE 9.5:
World's Simplest Motor

WORLD'S SIMPLEST MOTOR

Using a very basic motor (Figure 9.5), we examine
a simple example of electromagnetic induction
and the structure of a direct current motor.

Procedure

1. Assemble the motor according to the instructions.

2. Give the coil a little push to get it started.

Questions

• What happens? Why?

• Why doesn't the action stop?

▼
SAFETY NOTES

• Wear safety glasses or
goggles.

• Use caution when
handling wires: They are
sharp and can cut skin.

▼
SAFETY NOTE

Wear safety glasses or
goggles.

LEVITRON

The Levitron's (Figure 9.6) magnetic field
is formed by the toroidal (doughnut-
shaped) magnet inside of it. The physics
behind the Levitron is complicated, and
a good knowledge of physics is needed
to explain how this device works.

FIGURE 9.6: Levitron

Procedure

1. Adjust the Levitron so that the platform
    is precisely horizontal.

2. Place the transparent layer on the
    Levitron platform.

3. Spin the top on the transparent layer
    and slowly lift the layer about 5 cm
    above the level of the platform.

4. Slowly lower the transparent layer to
    the platform, leaving the top levitating.

    •   You will need to adjust the mass of
        the top through repeated trials to
        get it to levitate.

    •   If the top falls directly after being
        raised, it is too heavy.

    •   If the spinning top flies off, the
        top is too light or the Levitron
        platform is not level.

5. When you get the top to stay levitated,
    do the following tests:

    •   Move items over, under, and
        around the top as it rotates (e.g., a
        sheet of paper or your fingers).

    •   Place a drinking glass around the
        spinning top.

    •   Pretend to cut the air above the
        top with a pair of metal scissors.

Questions

•   Why does the spinning top stay
    levitated?

•   Why do eddy currents (based on
    Lenz's law) not stop the top?

•   What happens when you cut the air
    with metal scissors? Why?

Analysis

POINT THE WAY

In order to appreciate how the Magnetic Field Model (Figure 9.7) works, some basic facts of magnetism need to be understood.

How Is Magnetism Generated?

An electron's "spin" describes a type of internal rotation or angular momentum. The electrons in an atom effectively create a current loop around the atomic nucleus. The loops created by the electrons in an atom, combined with the electron's spin, create a magnetic moment. The magnetic moment is the basic source of the magnetic field.

Magnetic Fields

A magnetic moment (also called a magnetic dipole moment) is caused by the movement of electrical charges, such as an electric current or the internal angular momentum of an atom. The magnetic moment is a vector quantity. When influenced by an external magnetic field, the magnetic dipole moment (a vector) tends to turn parallel to the external magnetic field.

In the atoms of nonmagnetic materials, the magnetic moments of atoms are randomly oriented and tend to cancel each other out. In contrast, in magnetic objects the magnetic moments are not randomly oriented and therefore create a nonzero magnetic dipole moment. The north and south poles of a magnet are not separate magnetic units; rather, the poles are describing the direction of the overall magnetic moment and defining the direction of the magnetic field vector and whether the magnetic force is attractive or repulsive.

Different materials behave differently in an external magnetic field. Substances can be

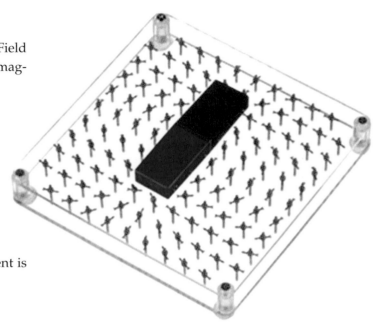

FIGURE 9.7: Magnetic Field Model

classified according to their responses to the external magnetic field as follows:

Diamagnetism

While interacting with an external magnetic field, the orbital motion of the electrons that cause a magnetic moment will change slightly. Lenz's law of induction describes how a current is created such that its effect is to magnetically oppose any change caused by the external magnetic field. Electrons in atoms can be thought of as loops of electrical current. An external magnetic field changes the electrons' orbital motion and the resulting magnetic moment. This change creates a magnetic field that is opposite in direction to the external magnetic field and opposes the external magnetic field. This effect is called diamagnetism and occurs in varying amounts in all atoms.

Paramagnetism

Magnetic moments caused by the spin and orbital motions of electrons in atoms usually cancel each other out, but in some materials the atoms' magnetic moments do not sum to zero. In this case, the external magnetic field will tend to rotate the magnetic moment vectors so that they strengthen the external magnetic field. Materials magnetized in this way are called paramagnetic materials. Paramagnetic substances can themselves become magnetized and strengthen the external magnetic field.

Ferromagnetism

In paramagnetic materials, the atoms' magnetic moments are not much affected by the surrounding atoms. However, in ferromagnetic materials the magnetic moments in atoms and molecules do interact with one another. Because of the material's structure, the magnetic moments are arranged by regions where they are parallel and reinforce one another; these regions are called magnetic domains. In a ferromagnetic material, an external magnetic field strengthens the magnetic domains that have magnetic moments in the same direction as the external magnetic field and weaken the others. The material is thus strongly magnetized; that is, it becomes a magnet.

The magnetic field model in this experiment has little pieces of metal in it—actually, small compass needles. The metal needles turn parallel to align with the magnetic field vectors of the external magnet. They are made from a ferromagnetic material.

FIGURE 9.8: Compasses around a bar magnet

MAGNETS IN MOTION

In this exploration, a number of compasses are arranged around a bar magnet. The compass needles turn so as to be parallel to the external magnetic field lines (Figure 9.8). The north ends of the little magnetic compass needles point toward the south magnetic pole of the bar magnet (or parallel to magnetic field lines that lead to the south pole). In the Earth's magnetic field, the compass needles point toward the Earth's magnetic south pole, which is located near the geographic north pole.

The moving charge creates (in addition to an electric field) a magnetic field. With an electric current, the direction of the magnetic field can be observed by the right hand rule: when the fingers of your right hand are curled around and the thumb is extended in the direction of the current, the curled fingers indicate the magnetic field direction (which circles around the wire attached to the battery), as shown in Figure 9.9.

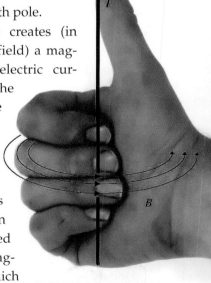

FIGURE 9.9: The right-hand rule shows the direction of the current and the magnetic field

FIGURE 9.10:
3D Magnetic Compass

WORLD'S SIMPLEST MOTOR

In this exploration, you examine the World's Simplest Motor (Figure 9.11), which is a basic direct-current motor. In the base of the motor, there is a battery and connectors where the coil is supported. The magnet on the platform creates a magnetic field that goes through the coil. The wire that the coil is made out of is coated with an insulator. The insulator is scratched off from only one side of the wire so that for half a rotation it is a conductor, and for half a rotation it is an insulator. The electric current creates a magnetic field in accordance with the right-hand rule (Figure 9.12, p. 128). When you curl your fingers in the direction of the current, your thumb points in the direction of the magnetic field. The magnetic field from the coil interacts with the magnetic field from the permanent magnet on the platform, and an attractive or repulsive force is created (depending on the direction of the current).

ALL ENCOMPASSING

The 3D Magnetic Compass (Figure 9.10) tests the strength of a magnetic field. The strength of a magnetic field can be described by the magnetic flux density, B. A current through a wire produces a magnetic flux density given by

$$B = \frac{\mu_0 I}{2\pi r}$$

where μ_0 is the permeability of a vacuum, I is the current, and r is the distance from the wire. The important part here is that the magnetic field intensity is directly proportional to the magnitude of the electric current and inversely proportional to the distance from the wire.

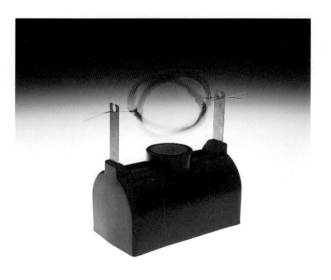

FIGURE 9.11: World's Simplest Motor

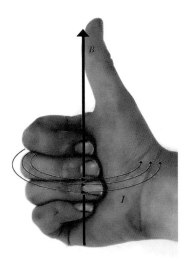

FIGURE 9.12: When you curl your fingers in the direction of the current, your thumb points in the direction of the magnetic field.

When the coil has turned halfway around, the insulated half of the wire touches the conductor, and the electric current vanishes. If the current didn't stop, the magnetic field would reverse direction, and the coil would be slowed down. The insulation on the half of the wire stopping the current is the reason why the magnetic interaction does not slow the coil down after half a rotation.

LEVITRON

Here, you spin a top and then watch as (after a certain amount of tinkering) it levitates above a platform. When the top is spinning above the magnet in the Levitron's base (Figure 9.13), the top is affected by magnetic repulsion, gravity, and air resistance. The magnetic repulsion is what keeps the top in the air, while the spinning continues due to the conservation of angular momentum. The top is a ceramic magnet, so eddy currents formed from the rotation are relatively small. (We will talk more about eddy currents in Chapter 10.) The most significant force to slow down the top is air resistance.

Scissors are usually made of aluminum, which is a nonmagnetic material, or from a ferromagnetic material, which is magnetized because of an external magnetic field. Eddy currents generated in nonferromagnetic scissors are not strong enough to disturb the top while it's spinning. Scissors made from a ferromagnetic material will be magnetized and will cause enough of a disturbance in the Levitron's magnetic field to make the top fall because of gravity.

FIGURE 9.13: Levitron

Web Resources

Explore the interactions between a compass and bar magnet, and then add the Earth to see how that affects the magnetic field.

http://phet.colorado.edu/en/simulation/magnet-and-compass

Use a battery and wire to make a magnet. See if you can change the strength or reverse the magnetic field.

http://phet.colorado.edu/en/simulation/magnets-and-electromagnets

Tutorials, demonstrations, and information about electricity and magnetism.

www.magnet.fsu.edu/education/tutorials/index.html

Visualize a magnetic field with iron filings.

http://solar-center.stanford.edu/magnetism/magneticfields.html

Relevant Standards

Note: The Next Generation Science Standards *can be viewed online at* www.nextgenscience.org/next-generation-science-standards.

PERFORMANCE EXPECTATIONS

HS-PS3-5

Develop and use a model of two objects interacting through electric or magnetic fields to illustrate the forces between objects and the changes in energy of the objects due to the interaction. [Clarification Statement: Examples of models could include drawings, diagrams, and texts, such as drawings of what happens when two charges of opposite polarity are near each other.]

HS-PS2-5

Plan and conduct an investigation to provide evidence that an electric current can produce a magnetic field and that a changing magnetic field can produce an electric current.

DISCIPLINARY CORE IDEAS

PS2.B: Types of Interactions

- Forces at a distance are explained by fields (gravitational, electric, and magnetic) permeating space that can transfer energy through space. Magnets or electric currents cause magnetic fields; electric charges or changing magnetic fields cause electric fields.

ESS2.A: Earth Materials and Systems

- Evidence from deep probes and seismic waves, reconstructions of historical changes in Earth's surface and its magnetic field, and an understanding of physical and chemical processes lead to a model of Earth with a hot but solid inner core, a liquid outer core, a solid mantle and crust.

10

ELECTROMAGNETIC INDUCTION

I f you did the "World's Simplest Motor" exploration in Chapter 9, you made a motor that used an electric current in a coil of wire to create a magnetic field. In this chapter, you will investigate toys that do the opposite—that is, they use a magnetic field to create an electric current. This is called electromagnetic induction.

Electromagnetic induction is the basis for producing energy in our society. This phenomenon is used in wind generators, hydroelectric generators, and nuclear power plants. Electromagnetic induction is also used in metal detectors, induction brakes, and transformers.

If the magnetic flux through a coil changes, it will generate a voltage. In a closed circuit, the voltage creates an electric current. A current created this way is called an induction (or induced) current. An induction current creates a magnetic field with a direction that tries to oppose the changes in the external magnetic field. In other words, an induction current tries to prevent the effects of the magnetic field that created the induction current in the first place! This phenomenon is known as Lenz's law.

According to Faraday's law of induction, the electromotive force (voltage) produced, E, is given by

$$E = -\frac{\Delta\Phi}{\Delta t}$$

where $\Delta\Phi$ is the change in magnetic flux, and Δt is the change in time.

The electrical currents produced, known as eddy currents, are created in a conductor when an external magnetic field around a conductor changes with time. Eddy currents are used in many applications, including induction ovens, electric power consumption meters, and metal detectors.

Relevant Standards

Note: The Next Generation Science Standards *can be viewed online at* www.nextgenscience.org/next-generation-science-standards.

PERFORMANCE EXPECTATIONS

HS-PS3-3

Design, build, and refine a device that works within given constraints to convert one form of energy into another form of energy. [Clarification Statement: Emphasis is on both qualitative and quantitative evaluations of devices. Examples of devices could include Rube Goldberg devices, wind turbines, solar cells, solar ovens, and generators....]

HS-PS3-5

Develop and use a model of two objects interacting through electric or magnetic fields to illustrate the forces between objects and the changes in energy of the objects due to the interaction. [Clarification Statement: Examples of models could include drawings, diagrams, and texts, such as drawings of what happens when two charges of opposite polarity are near each other.]

HS-PS2-5

Plan and conduct an investigation to provide evidence that an electric current can produce a magnetic field and that a changing magnetic field can produce an electric current.

DISCIPLINARY CORE IDEAS

PS2.B: Types of Interactions

- Forces at a distance are explained by fields (gravitational, electric, and magnetic) permeating space that can transfer energy through space. Magnets or electric currents cause magnetic fields; electric charges or changing magnetic fields cause electric fields.

MORE FUN

H ere are gadgets that we found very interesting and illustrative but that did not fit into the other chapters. However, many of these gadgets complement the explorations in earlier chapters.

Exploration

Hewitt-drewit

FIGURE 11.1: Bernoulli's Bag

SAFETY NOTES

Wear safety glasses or goggles.

WIND BAG

With the wind bag (often sold as Bernoulli's Bag; Figure 11.1), you'll see an interesting phenomenon that allows you to inflate a bag much faster than you might expect.

Procedure

1. Take two to four wind bags and tie a knot in one end of the tubes.

2. Place the bags next to each other on the floor. One should be kept aside for the referee.

3. Have a competition to inflate the bags—the student who does it fastest wins. One student can be the referee.

4. Next, the referee should inflate the bag (see the analysis for directions).

Question

*   What did you do with the bag when you performed the experiment? What happened?

*   What happens when the referee fills the bag? Why?

FIGURE 11.2: Doppler effect

DOPPLER BALL

The Doppler Ball can be used when studying the Doppler effect (Figure 11.2) and to illustrate why, for example, the sound from a car changes as it passes you.

Procedure

1. Put the battery into the sound source, and put the sound source in the ball.

2. Throw the ball to a friend on the other side of the classroom. Ask others to listen to how the sound is different when the ball is in motion. Then ask someone to throw the ball back.

3. Next, put the ball in a plastic bag and twirl the ball around yourself. Listen.

Question

*   Why is the sound different when the ball is moving?

▼
SAFETY NOTES

*   Wear safety glasses or goggles.

*   Make sure the trajectory path of the ball is clear of observers, fragile equipment, and so on, to prevent damage or injury.

*   Do not hold the sound source close to your ears.

MIRAGE

The name Mirage (Figure 11.3) describes the toy quite well. With the Mirage you can study images created in mirrors and discover the difference between a virtual and a real image.

Procedure

1. Put a small object in the middle of the lower mirror.

2. Put the lid on the Mirage.

3. Choose a spot where you can see the image on the lid. Viewing from some angles works better than other angles.

4. Ask the viewers to grab the object on the Mirage.

5. *Hint*: You can also point to the image with a laser. Be careful when doing this!

Questions

- Explain the concepts (a) virtual image and (b) real image.

- Is the image in the Mirage virtual or real? Why?

- What might a practical use of this phenomenon be?

FUN FLY STICK

The Fun Fly Stick (Figure 11.4) is like a small Van de Graaff generator. The generator of the Fun Fly Stick creates a positive electric charge on the outer shell of the stick. The outer shell is made of paper, which is an insulator. After you charge the stick, you can perform many static electricity demonstrations.

Procedure

1. Carefully take out one of the aluminum foil figures that came with the Fun Fly Stick. Put it in one hand. Turn the apparatus on by pushing the button, and charge the foil by touching it with the wand. Carefully shake the Fun Fly Stick to detach the foil figure. After you have charged the foil, you can make it fly as you wish.

2. Next, take the foil figure shaped like a butterfly. Charge it and make it fly. Put your hand above the flying butterfly so that the butterfly is between your hand and the Fun Fly Stick. Press the button and make the butterfly bounce between your hand and the Fun Fly Stick.

3. Take a sheet of paper, rub the Fun Fly Stick on it against a wall (as though you were ironing it), and make it stick to the wall. (This works better on some walls than others.)

▼
SAFETY NOTE

Wear safety glasses or goggles.

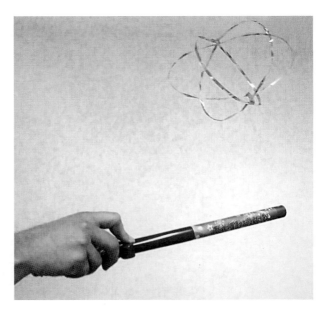

FIGURE 11.4: Fun Fly Stick

Questions

*   How does the Fun Fly Stick work?

*   Can you get an electric shock from the Fun Fly Stick? Why or why not?

*   What happens to the aluminum foil figure when it is charged? Why?

*   What makes the figure go up in the air?

*   Why does the butterfly bounce between your hand and the Fun Fly Stick?

*   Why does the paper stick to the wall?

FIGURE 11.5: Energy Ball

ENERGY BALL

The Energy Ball (Figure 11.5) lets you experience some concepts of open and closed electrical circuits.

Procedure

1. Make a circle of students in the classroom. Students must touch each other (e.g., by holding hands or touching elbows).

2. Put the Energy Ball between two students in the circle so that each student touches one of the two metal strips on the Energy Ball.

3. Ask one student in the circle to open the circuit.

Questions

- Explain the concepts (a) open electrical circuit and (b) closed electrical circuit.

- How does the Energy Ball work?

FLYING PIG

You can use the Flying Pig (Figure 11.6) to study uniform circular motion.

Procedure

1. Attach the string for the Flying Pig to the ceiling. Hang the pig from the string, and turn the power on.

2. Give the pig an appropriate initial velocity so it moves around in a circle.

3. Think about how to determine the amount of tension in the string.

Question

*   What needs to be measured in order to solve for the string tension force?

FIGURE 11.6: Flying Pig

▼
SAFETY NOTES

*   Wear safety glasses or goggles.

*   Do not hang the Flying Pig from a fire sprinkler head.

FIGURE 11.7: Bernoulli's Bag

WIND BAG

After the students have had their competition with the wind bags (Bernoulli's Bag; Figure 11.7), it's the referee's turn. Hold the bag wide open about 20–30 centimeters from your mouth and blow vigorously once or twice. This will be enough to fill the bag.

Some say that, based on the Bernoulli principle, which applies to fluids, moving air has a lower pressure than stationary air. However, there's a simple experiment you can do to show that simply blowing air out from your lungs does not produce a stream with a lower pressure than the surrounding air. (Note: This is most easily done while standing up.) Hold a piece of paper so that it hangs down vertically, and blow down on one side of the paper (Figure 11.8). Notice that the paper does not deflect toward the moving air. In an open environment, any pressure difference goes to zero in practically no time.

However, blowing out air can produce other effects. One of them is called entrainment. Entrainment occurs when the surrounding air gets pulled along with the air stream. Some fans work on this principle to blow out more air than you would otherwise expect. Filling the bag with just one blow can be explained by considering molecules and entrainment. The air molecules close to the stream are carried with the air molecules in the stream through entrainment. There is 10 times more air coming from outside the stream than in the stream, which causes the bag to fill with a single breath.

How is entrainment different from the Bernoulli effect? The Bernoulli effect involves changes of speed and pressure along flow lines in a tube or other confined enclosure. As mentioned above, in an open space any differences in air pressure would immediately push air from the higher-pressure region toward the lower pressure region, eliminating the original pressure difference between the moving and stationary air. Entrainment works on the microscopic level and is different from the Bernoulli effect.

Firefighters may use this phenomenon when they are removing smoke from buildings. Fans are placed at a distance from a door or window. This makes the rate of airflow much higher.

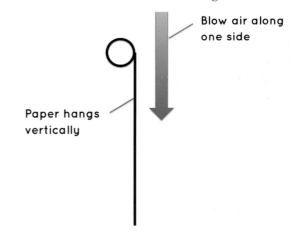

Blow air along one side

Paper hangs vertically

FIGURE 11.8: Blow down one side of the vertically hanging paper.

DOPPLER BALL

The Doppler Ball allows students to experience the Doppler effect (Figure 11.9) in a controlled environment. In general, if the source of the sound and the observer are both stationary (relative to each other), the sound doesn't change, and the frequency (and pitch) of the sound remains the same. If the observer and the source of the sound get closer to or farther from

FIGURE 11.9: Doppler effect

each another, the sound changes frequency. This phenomenon is called the Doppler effect. Sound travels through air as a longitudinal wave. As the source of the sound and the observer get closer together, the observer hears the sound higher in pitch, which means that the frequency of the sound is higher. This occurs because the observer "crashes" into crests of the wave more often. On the other hand, if the source and the observer move farther away from each other, the pitch of the sound heard is lower because the observer now crashes into wave crests less frequently.

This can be demonstrated with a wintry example. Two friends, Andy and Mandy, are having a snowball fight. Andy throws one snowball at Mandy every second from inside a snow fortress. After getting fed up with being hit by the snowballs at one-second intervals, Mandy starts running toward Andy. Now the balls seem to hit Mandy more often than once every second. From Mandy's perspective, it seems as if Andy is now throwing the balls at her more frequently than before.

Finally Mandy starts running away from Andy's snow fortress. Now she concludes that Andy is throwing the snowballs at her less frequently because the hits are spaced farther apart

from each other in time. According to Mandy, the frequency of the throws is lower.

The Doppler effect concept is used in a number of scientific fields. For example, the Doppler effect explains the phenomenon of redshift in astronomy. Astronomers use redshift to determine the speed at which astronomical objects are moving toward us or away from us (Figure 11.10). Redshift is similar to the Doppler effect with sound waves, only here we are talking about light waves. When an object such as a star moves away from the observer at a high velocity, the frequency of the light observed (like the frequency of snowball

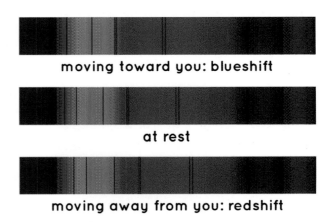

moving toward you: blueshift

at rest

moving away from you: redshift

FIGURE 11.10: Doppler-effect spectra

hits) is lower than the original frequency. Because the frequency and the wavelength are inversely proportional, the length of the wave is longer than it would be from a stationary star. The entire spectrum is shifted toward longer wavelengths. And since the longer wavelengths are at the red end of the spectrum, the shift of wavelengths is toward the red; hence the term redshift.

Finally, meteorologists use the Doppler effect for determining the speed of storms, and medical technicians can use the Doppler effect for determining the rate of blood flow through certain parts of your body and for determining the rate of a fetal heartbeat. Thus, the Doppler effect has many applications in the physical sciences and medicine for determining speeds.

MIRAGE

The image formed in the Mirage (Figure 11.11) is a real image because light rays are reflected from an object and actually reach the image on top of the Mirage. If you put the pig (which comes in the kit) on the lower mirror so that its nose is toward you, you'll notice that in the image the tail is toward you. Light rays reflected from the pig go from the upper mirror to the lower one, where they reflect to the eye of the viewer. A virtual image cannot be seen this way, and the real image could be reflected, for example, to a white screen or a wall.

FIGURE 11.11: Mirage

FUN FLY STICK

The Fun Fly Stick (Figure 11.12) shows that some materials can behave somewhat differently from normal when they come in contact with certain other materials. Some materials tend to donate electrons to other materials and are more likely to get positively charged, whereas other materials tend to take electrons when in contact and receive a negative charge. The *triboelectric series* can tell you if a material will charge negatively or positively.

The Fun Fly Stick consists of two cylinders, a rubber band, and a metal comb. The rubber band goes around the two cylinders, one of which is made of Teflon (PTFE) and the other of which is made of metal. The Teflon cylinder is charged negatively when the rubber band contacts the cylinder while it is rotating. Close to the cylinder, there is a metal comb so that the rubber band goes between the cylinder and the comb. The metal comb is grounded.

The cylinder's negative charges repel electrons in the metal comb, and therefore, the needles of the comb are charged positively and a strong electric field is created between the cylinder and the comb. The resulting electric field is so strong that it can ionize air molecules. When the air molecules are ionized, the positive ions move toward the Teflon cylinder and attach to the rubber band, which is traveling around the cylinder. The positive charge traveling with the rubber band ultimately gets transferred to a metal film that covers the Fun Fly Stick, so the metal film is charged positively. This metal film is insulated with paper, so you don't get an electric shock from the Fun Fly Stick. Despite the high voltages present, the resistance of the paper prevents the formation of large currents.

When the aluminum foil is in contact with the Fun Fly Stick, the free electrons transfer to the

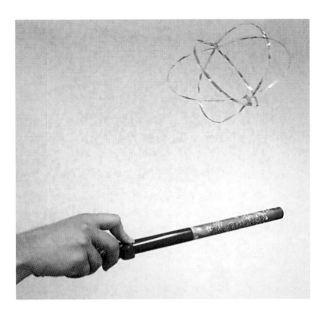

FIGURE 11.12: Fun Fly Stick

free electrons as charge carriers. Sliding the Fun Fly Stick along the paper can do two things to the paper: First, it can polarize some of the molecules in the paper. The molecules on the paper's surface tend to turn so that their negative poles are directed toward the positively charged Fun Fly Stick. Second, it can cause some electrons to move from the paper to the Fun Fly Stick, leaving the paper with a net charge. Both of these effects can contribute to the paper sticking to the wall. For the paper to stick, the wall, too, must develop a net charge in the vicinity of the charged paper. This might result from polarization of the molecules in the wall, or might result from induction, that is, charges flowing through the wall to the spot where the paper is (provided that the wall has at least a small bit of conductivity).

ENERGY BALL

The exploration with the Energy Ball (Figure 11.13) helps you investigate closed and open circuits by creating a human circuit. When the circuit is closed (i.e., everyone is holding hands or touching elbows), the ball makes noise and flashes a light.

positively charged stick, leaving the foil with an excess of positive charges. Because like charges repel each other, the aluminum foil opens into a large shape. In addition to the repulsion of charges, gravity and air currents affect the foil as it moves through the air.

In one step, you bounce the butterfly between your hand and the Fun Fly Stick. The bouncing happens because after the butterfly has been charged, the negative charges in your hand attract the foil butterfly. When the butterfly touches your hand, some electrons get transferred, and then the hand's contact point is positively charged so that the like charges repel each other and the foil butterfly moves away from your hand. The positively charged Fun Fly Stick then attracts the free electrons from the foil and the process starts over.

In the final demonstration of the power of static electricity, you stick a piece of paper to the wall. Paper is an insulator, so it does not have

FIGURE 11.13: Energy Ball

In a circuit, there is usually a source of voltage and current, resistors, and sometimes, capacitors or inductors. The voltage of the battery in the Energy Ball causes an electric current in a closed circuit; a closed circuit is a loop that enables the current to return. In an open circuit, there is no current present in spite of having voltage. A closed circuit can be made open (and vice versa) with a switch.

The operation of the Energy Ball relies on two different circuits. The Energy Ball has a channel field-effect transistor (FET) through which the current passes if the resistance between the metal strips is measurable. In the circuit made by students, there is a very small current. When that current is flowing, the transistor allows current in another circuit to flow through the lamp and speaker.

FLYING PIG

The Flying Pig (Figure 11.14) travels at a steady speed along a circular path. Using Newton's second law, we can analyze the forces separately in the vertical and horizontal directions. The sum of the forces in the vertical (y) direction is

$$\Sigma F_y = T_y + mg = 0,$$

where T_y is the vertical component of the string tension, and mg is the force of gravity (i.e., weight).

The sum of the forces in the horizontal (x) direction is

$$\Sigma F_x = T_x + ma_c = 0$$

FIGURE 11.14: Flying Pig

where a_c is the centripetal acceleration. Substituting $T_x = T\sin\alpha$ (the horizontal component of the tension), with α being the angle between the string and the vertical, and $a_c = v^2 / r$ for the centripetal acceleration, and solving for T, we find that the magnitude of the tension is given by

$$T = \frac{mv^2}{r \sin \alpha}$$

Therefore, to calculate the tension force in the string, we would need to measure the radius, angle, period of revolution, and mass of the pig.

Web Resources

A visualization of the Doppler effect.
http://lectureonline.cl.msu.edu/~mmp/applist/doppler/d.htm

More information about the Mirage device.
www.arborsci.com/media/datasheet/P2-7070_Additional.pdf

Build circuits and take measurements with an ammeter and voltmeter.
http://phet.colorado.edu/en/simulation/circuit-construction-kit-dc

An explanation of aerodynamic lift (and some comments about the wind bag near the end).
www.aerodynamiclift.com

Relevant Standards

Note: The Next Generation Science Standards *can be viewed online at* www.nextgenscience.org/next-generation-science-standards.

PERFORMANCE EXPECATIONS

Flying Pig

HS-PS2-1

Analyze data to support the claim that Newton's second law of motion describes the mathematical relationship among the net force on a macroscopic object, its mass, and its acceleration.

Energy Ball

HS-PS3-3

Design, build, and refine a device that works within given constraints to convert one form of energy into another form of energy. [Clarification Statement: Emphasis is on both qualitative and quantitative evaluations of devices. Examples of devices could include Rube Goldberg devices, wind turbines, solar cells, solar ovens, and generators....]

Fun Fly Stick

HS-PS1-3

Plan and conduct an investigation to gather evidence to compare the structure of substances at the bulk scale to infer the strength of electrical forces between particles.

HS-PS3-5

Develop and use a model of two objects interacting through electric or magnetic fields to illustrate the forces between objects and the changes in energy of the objects due to the interaction. [Clarification Statement: Examples of models could include drawings, diagrams, and texts, such as drawings of what happens when two charges of opposite polarity are near each other.]

Doppler Ball (redshift)

HS-ESS1-2

Construct an explanation of the Big Bang theory based on astronomical evidence of light spectra, motion of distant galaxies, and composition of matter in the universe. [Clarification Statement: Emphasis is on the astronomical evidence of the red shift of light from galaxies as an indication that the universe is currently expanding … .]

Fun Fly Stick, Bernoulli Bag

HS-PS2-6

Communicate scientific and technical information about why the molecular-level structure is important in the functioning of designed materials. [Clarification Statement: Emphasis is on the attractive and repulsive forces that determine the functioning of the material. Examples could include why electrically conductive materials are often made of metal.]

Mirage

HS-PS4-5

Communicate technical information about how some technological devices use the principles of wave behavior and wave interactions with matter to transmit and capture information and energy. [Clarification Statement: Examples could include solar cells capturing light and converting it to electricity; medical imaging … .]

DISCIPLINARY CORE IDEAS

PS2.B: Types of Interactions

- Forces at a distance are explained by fields (gravitational, electric, and magnetic) permeating space that can transfer energy through space. Magnets or electric currents cause magnetic fields; electric charges or changing magnetic fields cause electric fields.

APPENDIX

HOW TO ORDER THE GADGETS AND GIZMOS

Materials to support the lessons and experiments found throughout this book are available from Arbor Scientific in NSTA High School Physics Kit #1 and #2. Each kit includes about 25 different tools (listed below) to support lessons in angular momentum, buoyancy, magnetism, pressure, energy, and more.

Visit *arborsci.com* to order the kits or learn more about their contents.

KIT #1: PK-0100		KIT #2: PK-0110	
PRESSURE AND FORCE		**BUOYANCY**	
Atmospheric Mat Pressure Globe Hollow Prism	Super Bottle Rocket Launcher Atmospheric Pressure Cups	Solar Bag Boat & Rock Galileo's thermometer	Cartesian Diver Poly Density bottle
LAWS OF THERMODYNAMICS		**ANGULAR MOMENTUM**	
Reversible Thermoelectric Demo Elasticity of Gases Demo Fire Syringe	Drinking Bird Ice Melting Blocks	Rotating Platform Extreme Gyroscope IR-Controlled UFO	Perpetual Top Celt Power Ball with Speed Meter
ENERGY		**MAGNETISM**	
Introductory Energy and Motion Lab Colliding Steel Spheres Euler's Disc Happy / Unhappy Balls	Crookes Radiometer Magnetic Accelerator Dropper Popper Astroblaster	Magnetic Field Model Small Clear Compasses 3D Magnetic Compass	World's Simplest Motor Levitron w/ Starter
COLORS		**ELECTROMAGNETIC INDUCTION**	
RGB Snap Lights and Spinner Quantitative Spectroscope	Replacement Color Filters for 2009 Color Addition Spotlight Primary Color Light Sticks	Lenz's Law Apparatus Electromagnetic Flashlight	Deluxe Hand Crank Generator 1 Farad Capacitor
RESONANCE		**MORE FUN**	
Set of 8 Boomwhackers Sound Pipe Music Box Mechanism	Standing Wave Apparatus Singing Rods w/ Rosin	Mirage Doppler Ball Flying Pig	Fun Fly Stick Energy Ball Bernoulli's Bags
2D-MOTION			
Vertical Acceleration Demonstrator Ballistics Car Introductory Energy and Motion Lab	Air-Powered Projectile Angle Wedges Launch Pad		

CREDITS

IMAGES

MEDIAKETTU JARI PEURAJÄRVI

ARBOR SCIENTIFIC

RPM SPORTS

PAUL HEWITT

MATTI KORHONEN

OTHER CONTRIBUTIONS

OLIVIA BOBROWSKY

ANDREW GLOR

INDEX

*Page numbers in **boldface** type refer to figures.*

A

Absorption spectrum, 55
Adding and Subtracting Colors exploration, 53, **53**
   analysis of, **56**, 56–57, **57**
Adiabatic process, 10, 19, 24, 25
Air pressure explorations, 1
   It's a Hold-up!, 8
   Pressure Globe, 10
   Pressure Power, 2, 7
   Prism Pressure, 4
   Radiant Rotation, 42
   Water Rocket, 10
   Wing Bag, 150
Air resistance explorations
   Dancing Disc, 41
   Ejector Seat, 97
   Energy on Wheels, 40
   Launch and Land, 99
   Levitron, 128
   Magnetic Accelerator, 41
   Need for Speed, 98
All Encompassing exploration, 122, **122**
   analysis of, 127, **127**
Angular Momentum explorations, **105**, 105–117
   Perpetual Top, 110, **110**, 114, **114**
   Power Ball, 108, **108**, 112–113, **113**
   Propeller Puzzle, 107, **107**, 112, **112**
   Speed Spinning, 106, **106, 111**, 111–112, **112**
   standards addressed by, 116–117
   Web resources for, 115
   Wobbly Top, 109, **109, 113**, 113–114
Angular velocity explorations, 105
   Power Ball, 112–113
   Speed Spinning, 106, 111–112
   Wobbly Top, 113–114
Assessment, xv, xvi
Astroblaster exploration, 39, **39**
   analysis of, 44, **44**

B

Bernoulli effect, 150
Boomwhackers exploration, 65, **65**
   analysis of, 70, **70**
Boyle's law, 18, 23
Buoyancy explorations, **79**, 79–89
   Cartesian Diver, 80, **80**, 86, **86**
   Galileo's Thermometer, 81, **81**, 86, **86**
   Mix It Up, 82, **82**, 86, **86**
   Rock the Boat, 84–85, **85**, 87, **87**
   Solar Bag, 83, **83**, 86–87, **87**
   standards addressed by, 89
   Web resources for, 88

C

Carnot engine, 19, 24–25, **25**
Cartesian Diver exploration, 80, **80**
   analysis of, 86, **86**
Celt, 41
Chemical elements, spectral lines of, 55
Cohen, Jamie, xiv
Collision explorations
   elastic and inelastic collisions, 37
   Happy / Unhappy Balls, 37, 43
   Hot Shot, 34, 42
   Magnetic Accelerator, 36, 43
Color vision. *See* Visible Light and Colors explorations
Combining Colors exploration, 54, **54**
   analysis of, 57, **57**
Cone and rod cells in eye, 51
Conservation of angular momentum, 106, 107, 110, 112, 114
Conservation of energy, 31, 48, 99
Conservation of momentum, 12, 106, 107, 110, 112, 114, 116
Crank-a-Current exploration, 136, **136**
   analysis of, 138, **138**
Critical thinking, xiv

Crosscutting concepts
  for Buoyancy explorations, 89
  for Energy explorations, 49
  for Laws of Thermodynamics explorations, 29
  for Pressure and Force explorations, 13
  for Resonance explorations, 76
  for Two-Dimensional Motion explorations, 103
  for Visible Light and Colors explorations, 61

D
Dancing Disc exploration, 33, **33**
  analysis of, 41, **41**
Density. *See* Buoyancy explorations
Diamagnetism, 120, 125
Direct current motor, 123, **123, 127,** 127–128
Disciplinary core ideas
  for Electromagnetic Induction explorations, 140
  for Energy explorations, 48–49
  for Laws of Thermodynamics explorations, 29
  for Magnetism explorations, 130
  for More Fun explorations, 157
  for Pressure and Force explorations, 13
  for Resonance explorations, 75–76
  for Two-Dimensional Motion explorations, 103
  for Visible Light and Colors explorations,
    60–61
Doppler Ball exploration, 145, **145,** 157
  analysis of, **151,** 151–152
Doppler effect, 145, 151–152
Drinking Bird exploration, 20, **20**
  analysis of, 25, **25**
Dropper Popper exploration, 38, **38**
  analysis of, 44, **44**

E
Eddy currents, 124, 128, 133, 134, 137
Eddy Oddity exploration, 134, **134**
  analysis of, 137, **137**
Einstein, Albert, xii, xiii, xiv
Ejector Seat exploration, 93, **93**
  analysis of, 97, **97**
Elasticity of Gases Demo, 18, **18**
  analysis of, 23, **23**
Electrical circuits, 114, 121, 122, 148

Electrical energy, 21, 26, 31
Electromagnetic Induction explorations, 123,
    133–140
  Crank-a-Current, 136, **136,** 138, **138**
  Eddy Oddity, 134, **134,** 137, **137**
  equation for electromotive force, 133
  Shake & Shine, 135, **135,** 138, **138**
  standards addressed by, 140
  Web resources for, 139
Electromagnetic radiation, 59, 60–61
Emission spectrum, 55
Energy Ball exploration, 148, **148,** 156
  analysis of, **153,** 153–154
Energy explorations, 31–49
  Astroblaster, 39, **39,** 44, **44**
  conservation of energy, 31, 48
  Dancing Disc, 33, **33,** 41, **41**
  Dropper Popper, 38, **38,** 44, **44**
  Energy on Wheels, 32, **32,** 40, **40**
  equation for work, 40
  Happy / Unhappy Balls, 37, **37,** 43, **43**
  Hot Shot, 34, **34,** 42, **42**
  Magnetic Accelerator, 36, **36,** 43, **43**
  Radiant Rotation, 35, **35,** 42, **42**
  standards addressed by, 46–49
  types of energy, 31
  Web resources for, 45
Energy on Wheels exploration, 32, **32**
  analysis of, 40, **40**
Eye, light-sensitive cells of, 51

F
Faraday's law of induction, 133, 135, 137, 138
Ferromagnetism, 120, 126, 128
Finnish educational system, xvi
Fire Syringe exploration, 19, **19,** 19, **19**
  analysis of, 24–25, **24–25**
Floating and sinking. *See* Buoyancy explorations
Flying Pig exploration, 149, **149,** 156
  analysis of, 154, **154**
Force. *See* Pressure and Force explorations
Friction, 40, 41, 43, 71, 97, 110
Fun Fly Stick exploration, 147, **147,** 156, 157
  analysis of, 152–153, **153**
Fundamental frequency, 63, 65, 70, 72

G

Gadgets and gizmos, xiii, xvi
    for Angular Momentum explorations
        Extreme Gyroscope, **109, 113**
        IR-Controlled UFO Flyer, **107, 112**
        Perpetual Top, **110, 114**
        Power Ball with Speed Meter, **108, 113**
        Rotating Platform, **106, 111**
    for Buoyancy explorations
        Boat & Rock, **85, 87**
        Cartesian Diver, **80, 86**
        Galileo's thermometer, **81, 86**
        Poly Density Bottles, **82, 86**
        Solar Bag, **83, 87**
    for Electromagnetic Induction explorations
        Deluxe Hand Crank Generator, **136, 138**
        Electromagnetic Flashlight, **135, 138**
        Lenz's Law Apparatus, **134, 137**
    for Energy explorations
        Astroblaster, **39, 44**
        Car Track, **32, 40**
        Colliding Steel Spheres, **34, 42**
        Crookes Radiometer, **35, 42**
        Dropper Popper, **38, 44**
        Euler's Disc, **33, 41**
        Happy / Unhappy Balls, **37, 43**
        Magnetic Accelerator, **36, 43**
    for Laws of Thermodynamics explorations
        Drinking Bird, **20, 25**
        Elasticity of Gases demo, **18, 23**
        Fire Syringe, **19, 24**
        Ice Melting Blocks, **16, 22**
        Reversible Thermoelectric Demo, **21, 26**
        temperature sensors, **17**
    for Magnetism explorations
        Levitron, **124, 128**
        Magnetic Field Model, **120, 125**
        Small Clear Compass, **121, 126**
        3D Magnetic Compass, **122, 127**
        World's Simplest Motor, **123, 127**
    for More Fun explorations
        Bernoulli's Bag, **144, 150**
        Doppler Ball, **145, 151**
        Energy Ball, **148, 153**
        Flying Pig, **149, 154**
        Fun Fly Stick, **147, 153**
        Mirage, **146, 152**
    ordering kits of, 159
    for Pressure and Force explorations
        Atmospheric Mat, **3, 8**
        Atmospheric Pressure Cups, **2, 7**
        Hollow Prism, **4, 9**
        Pressure Globe, **5, 10**
        Super Bottle Rocket Launcher, **6, 10**
    for Resonance explorations
        Boomwhackers, **65, 70**
        Music Box, **67, 71**
        Singing Rod, **66, 71**
        Sound Pipe, **68, 72**
        Standing Wave Apparatus, **64, 69**
    for Two-Dimensional Motion explorations
        Air-Powered Projectile, **96, 99**
        Ballistics Car, **93, 97**
        Car Track, **95, 98**
        Vertical Acceleration Demonstrator, **92, 97**
    for Visible Light and Colors explorations
        color filters, **53, 56, 57**
        Quantitative Spectroscope, **52, 55**
        RGB Snap Lights and Spinner, **54, 57**
Galileo's Thermometer exploration, 81, **81**
    analysis of, 86, **86**
Gas pressure, 4, 7, 18, 25
Generators, 133, 140, 156
    hand crank, 136, 138
    Van de Graaff, 147
Group activities and discussions, xiii, xv–xvi
Gyroscope, 109, **109, 113**, 113–114, **114**

H

Hands-on activities, xiii, xvii, 1
Happy/Unhappy Balls exploration, 37, **37**
    analysis of, 43, **43**
Harmonic number, 63, 65, 70, 71, 72
Heat conduction, 16, 22
Hot Shot exploration, 34, **34**
    analysis of, 42, **42**
Hydrostatic pressure, 4, 9

I

Inertia, 41, 105, 112, 114

Inquiry-based instruction, xiii, xvi
Introductory Energy and Motion Lab, 32, **32**
    analysis of, 40, **40**
Isotherm, 23, **23**
It's a Hold-up! exploration, 3, **3**
    analysis of, 8, **8**

K
Kinetic energy, 25, 31, 48, 101
    during collisions, 34, 36, 37, 42, 43, 44
    conversion of potential energy to, 32, 40, 43, 44
    conversion of thermal energy to, 21, 42
    rotational, 31, 33
    temperature and, 24, 29
    translational, 31, 33
Kits of gadgets and gizmos, 159

L
Launch and Land exploration, 96, **96**
    analysis of, 99, **99**
Laws of Thermodynamics explorations, 15–29
    Drinking Bird, 20, **20**, 25, **25**
    Fire Syringe, 19, **19**, 24–25, **24–25**
    Meltdown, 16–17, **16–17**, 22, **22**
    Pressure Plunger, 18, **18**, 23, **23**
    Running Hot and Cold, 21, **21**, 26, **26**
    standards addressed by, 28–29
    thermodynamic equilibrium, 16
    Web resources for, 27
Lenz's law, 124, 125, 133, 134, 137
Levitron exploration, 124, **124**
    analysis of, 128, **128**
Light. *See* Visible Light and Colors explorations
Light mill, 42
Linear momentum, 105, 111
Lodestone, 119

M
Magnetic Accelerator exploration, 36, **36**
    analysis of, 43, **43**
Magnetism explorations, 119–130
    All Encompassing, 122, **122**, 127, **127**
    discovery and uses of magnetism, 119
    Levitron, 124, **124**, 128, **128**
    magnetic poles, 119,121,126
    Magnets in Motion, 121, **121**, 126, **126**
    Point the Way, 120, **120**, **125**, 125–126
    standards addressed by, 130
    Web resources for, 129
    World's Simplest Motor, 123, **123**, **127**, 127–128, **128**, 133
Magnets in Motion exploration, 121, **121**
    analysis of, 126, **126**
Mechanical energy, 15, 32
Meltdown exploration, 16–17, **16–17**
    analysis of, 22, **22**
Mirage exploration, 146, **146**
    analysis of, 152, **152**
Misconceptions of students, xiv, xv
Mix It Up exploration, 82, **82**
    analysis of, 86, **86**
More Fun explorations, 143–157
    Doppler Ball, 145, **145**, **151**, 151–152
    Energy Ball, 148, **148**, **153**, 153–154
    Flying Pig, 149, **149**, 154, **154**
    Fun Fly Stick, 147, **147**, 152–153, **153**
    Mirage, 146, **146**, 152, **152**
    standards addressed by, 156–157
    Web resources for, 155
    Wind Bag, 144, **144**, 150, **150**
Music Box exploration, 67, **67**
    analysis of, 71, **71**
Musical instruments. *See* Resonance explorations

N
Nature of science connections
    for Angular Momentum explorations, 117
    for Buoyancy explorations, 89
    for Energy explorations, 47–48
    for Laws of Thermodynamics explorations, 28
    for Pressure and Force explorations, 12
    for Resonance explorations, 75
    for Two-Dimensional Motion explorations, 103
    for Visible Light and Colors explorations, 60
Need for Speed exploration, 94–95, **95**
    analysis of, 98, **98**

Newton, Isaac, xiv, 57
Newton's laws, 103, 112, 114, 154, 156
Next Generation Science Standards (NGSS), xiv
    for Angular Momentum explorations, 116–117
    for Buoyancy explorations, 89
    for Electromagnetic Induction explorations, 140
    for Energy explorations, 46–49
    for Laws of Thermodynamics explorations, 28–29
    for Magnetism explorations, 130
    for More Fun explorations, 156–157
    for Pressure and Force explorations, 12–13
    for Resonance explorations, 74–76
    for Two-Dimensional Motion explorations, 101–103
    for Visible Light and Colors explorations, 59–61

O
Ordering gadgets and gizmos, 159
Oscillation. *See* Resonance explorations

P
Paramagnetism, 120, 126
Peer instruction, xiii
Peltier effect, 26
Performance expectations
    for Angular Momentum explorations, 116
    for Electromagnetic Induction explorations, 140
    for Energy explorations, 46
    for Laws of Thermodynamics explorations, 28
    for Magnetism explorations, 130
    for More Fun explorations, 156–157
    for Pressure and Force explorations, 12
    for Resonance explorations, 74
    for Two-Dimensional Motion explorations, 101
    for Visible Light and Colors explorations, 59
Perpetual Top exploration, 110, **110**
    analysis of, 114, **114**
Phenomenon-based learning (PBL), xiii–xvii
    definition of, xiii
    difference from project- or problem-based
        learning, xiv
    in Finland, xvi
    objective of, xiv
    use of gadgets and gizmos in, xiii, xvi
Photons, 31, 61
Pitch, 65, 66, 68, 70, 71
Point the Way exploration, 120, **120**
    analysis of, **125,** 125–126
Potential energy, 31
    conversion to kinetic energy, 32, 40, 43, 44
    effect of magnets on, 36, 43
    gravitational, 31, 44
    temperature and, 29
Power Ball exploration, 108, **108**
    analysis of, 112–113, **113**
Pressure and Force explorations, 1–13
    equation for pressure, 7, 8
    It's a Hold-up!, 3, **3,** 8, **8**
    Pressure Globe, 5, **5,** 10, **10**
    Pressure Power, xv, 2, 7, **7**
    Prism Pressure, 4, **4,** 9, **9**
    standards addressed by, 12–13`
    Water Rocket, 6, **6,** 10, **10**
    Web resources for, 11
Pressure Globe exploration, 5, **5**
    analysis of, 10, **10**
Pressure Plunger exploration, 18, **18**
    analysis of, 23, **23**
Pressure Power exploration, xv, 2, **2**
    analysis of, 7, **7**
Pressure–volume *(P-V)* relationship, 18, 23, **23**
Pressure–volume–temperature relationship, 19, 24, **24**
Prism Pressure exploration, 4, **4**
    analysis of, 9, **9**
Programme of International Student Assessment (PISA), xvi
Propeller Puzzle exploration, 107, **107**
    analysis of, 112, **112**

R
Race to the Bottom exploration, 92, **92**
    analysis of, 97, **97**
Radiant Rotation exploration, 35, **35**
    analysis of, 42, **42**

INDEX

Radiation pressure, 35
Rattleback, 41
Relative humidity, 10
Resonance explorations, 63–76
    Boomwhackers, 65, **65,** 70, **70**
    Music Box, 67, **67,** 71, **71**
    Singing Rods, 66, **66,** 71, **71**
    Sound Pipe, 68, **68,** 72, **72**
    standards addressed by, 74–76
    Standing Wave, 64, **64,** 69, **69**
    Web resources for, 73
Reversible Thermoelectric Demo, 21, **21**
    analysis of, 26, **26**
Rock the Boat exploration, **84,** 84–85, **85**
    analysis of, 87, **87**
Rod and cone cells in eye, 51
Running Hot and Cold exploration, 21, **21**
    analysis of, 26, **26**

S
Safety acknowledgment form, xvii
Safety notes, xv. *See also specific explorations*
Science and engineering practices
    for Angular Momentum explorations, 116–117
    for Buoyancy explorations, 89
    for Energy explorations, 46–47
    for Laws of Thermodynamics explorations, 28
    for Pressure and Force explorations, 12
    for Resonance explorations, 74–75
    for Two-Dimensional Motion explorations, 101–102
    for Visible Light and Colors explorations, 59–60
Science journals, xv
Science literacy, xvi
Seebeck effect, 26
Shake & Shine exploration, 135, **135**
    analysis of, 138, **138**
Singing Rods exploration, 66, **66**
    analysis of, 71, **71**
Solar Bag exploration, 83, **83**
    analysis of, 86–87, **87**
Sound energy, 31, 34, 40, 41, 42, 43, 48
Sound Pipe exploration, 68, **68**
    analysis of, 72, **72**

Sound waves. *See* Resonance explorations
Spectroscope exploration, 52, **52**
    analysis of, 55, **55**
Speed of sound, 66, 70
Speed Spinning exploration, 106, **106**
    analysis of, **111,** 111–112, **112**
Standing Wave exploration, 64, **64**
    analysis of, 69, **69**
Static electricity, 147, 153
Surface tension of water, 9

T
Temperature, 29
Temperature explorations
    Drinking Bird, 20, 25
    Fire Syringe, 19, 24–25
    Meltdown, 16–17, 22
    Pressure Plunger, 18, 23
    Running Hot and Cold, 21, 26
    Water Rocket, 10
Thermal energy, 28, 29, 31, 48, 49, 101
    in Carnot engine, 25
    conversion of electromagnetic radiation to, 61
    conversion to electrical energy, 21, 26
    conversion to kinetic energy, 21, 42
    temperature and, 29
    total, 29
    transfer in metals, 22
Thermal engine, 21
Thermal insulators, 22
Thermodynamics. *See* Laws of Thermodynamics explorations
Thinking like scientists, xv–xvi
Time required for explorations, xv
Torque, 41, 108, 112–114
Two-Dimensional (2D) Motion explorations, 91–103
    Ejector Seat, 93, **93,** 97, **97**
    Launch and Land, 96, **96,** 99, **99**
    Need for Speed, 94–95, **95,** 98, **98**
    Race to the Bottom, 92, **92,** 97, **97**
    standards addressed by, 101–103
    Web resources for, 100

U
Ultraviolet light, 55, 61

V
Velocity explorations, 91
    angular velocity, 105
        Power Ball, 112–113
        Speed Spinning, 106, 111–112
        Wobbly Top, 113–114
    Astroblaster, 44
    Dancing Disc, 41
    Ejector Seat, 93, 97
    Energy on Wheels, 32, 40
    Flying Pig, 149
    Launch and Land, 96, 99
    Need for Speed, 94–95, 98
    Race to the Bottom, 97
Visible Light and Colors explorations, 51–61
    Adding and Subtracting Colors, 53, **53, 56,**
        56–57, **57**
    Combining Colors, 54, **54,** 57, **57**

    rod and cone cells in eye, 51
    Spectroscope, 52, **52,** 55, **55**
    standards addressed by, 59–61
    Web resources for, 58
Voltage explorations, 133
    Crank-a-Current, 136, 138
    Eddy Oddity, 134, 137
    Energy Ball, 154
    Fun Fly Stick, 152
    Running Hot and Cold, 26
    Shake & Shine, 135, 138

W
Water Rocket exploration, 6, **6**
    analysis of, 10, **10**
Wind Bag exploration, 144, **144**
    analysis of, 150, **150**
Wobbly Top exploration, 109, **109**
    analysis of, **113,** 113–114
World's Simplest Motor exploration, 123, **123,** 133
    analysis of, **127,** 127–128, **128**